Ham Radio
Bible

A Comprehensive Guide to Ham Radio Mastery for
Navigating the Frequencies of Communication,
From Novice to Expert

McBunny Albert

TABLE OF CONTENTS

INTRODUCTION TO HAM RADIO

The hobby and service known as amateur radio, often known as ham radio, is a popular activity that brings together people, electronics, and communication. Without the Internet or mobile phones, people may communicate using ham radio to converse across town, over the globe, or even into space. It's not only entertaining and sociable, but it's also informative and may be a lifeline in times of need. Amateur radio operators are involved for a variety of reasons; however, they all share a fundamental understanding of radio technology and operating principles. Additionally, they must pass an examination to obtain a license from the Federal Communications Commission (FCC) to operate on radio frequencies that are referred to as the "**Amateur Bands.**" These bands are radio frequencies that have been designated by the FCC for use by ham radio operators.

The term "**ham radio**" refers to a specific sort of communication equipment that operates on certain radio frequencies that have been assigned by regulatory organizations within the government. At the beginning of the 20th century, when amateur radio operators were seen as "**amateurish**" by commercial radio operators, the word "**ham**" is said to have arisen. The competence and commitment of ham radio operators, on the other hand, have been shown throughout time. Individuals are allowed to investigate and experiment with radio communication via the use of ham radio, which is the major motive for its existence. Ham radio operators interact with one another using a wide range of modes, including voice, Morse code, digital modes, and many more. The pastime has a significant social component, but it also plays important roles in public service, disaster relief, and as a method of self-sufficiency in times of emergencies.

CHAPTER ONE
THE WORLD OF AMATEUR RADIO

Overview

Chapter One introduces us to the world of Ham Radio. Here, you will learn its history, evolution, its role, licensing and regulations, its types and so much more.

History and Evolution of Ham Radio

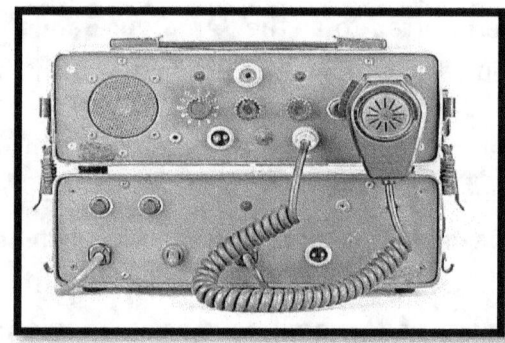

Amateur radio has a long and illustrious history that dates back more than a century. During the early part of the 20th century, pioneers such as Guglielmo Marconi and Nikola Tesla were responsible for the creation of wireless telegraph technology. To begin with, amateur radio was largely used to conduct experiments; nevertheless, it rapidly gained popularity among enthusiasts who were interested in communicating via the use of radio waves. There have been millions of individuals all over the globe who have become licensed HAM radio operators since the first Amateur Radio license was awarded in 1912 in the United States of America. Even in modern times, Ham Radio continues to be a well-liked activity that is embraced by individuals of all ages and walks of life. The provision of this method of emergency communication during times of calamity, when conventional modes of communication may be unavailable or unreliable, is made possible by it. Additionally, Ham Radio provides an opportunity for experimenting with new technologies and methods that are associated with wireless communications. It makes it possible for persons who have similar interests in this one-of-a-kind sort of communication technology to engage in social contact with one another.

Ham Radio, which is often referred to as amateur radio, is a kind of radio communication that enables people to communicate with one another via the use of radio frequencies that have been set aside specifically for that purpose. To get a license from the regulatory authorities of their various countries, Ham Radio operators are required to show that they are familiar with the necessary laws and regulations by passing an examination. When compared to professional radio operators, amateur radio operators were believed to be less talented and less experienced. The name "**ham**" was formerly used to describe these individuals.

Origins: Marconi and Beyond

Many people consider Guglielmo Marconi to be the "father" of modern radio communication. He is credited with developing the first radio system that could be used in everyday life in the year 1895. The contributions made by Marconi made it possible for amateur radio enthusiasts to investigate and experiment with the possibilities offered by wireless communication technology. Did you know that the word "**ham**" was first used to describe these amateur operators, who were often considered to have a lower level of expertise and experience compared to their professional counterparts? Despite the widespread belief to the contrary, ham radio operators were instrumental in the invention and growth of wireless communication technologies during the early 20th century. Ham radios were used for a variety of purposes, including ship-to-shore communications, emergency announcements during natural catastrophes, and the connection of individuals who were separated by enormous distances. It was also a crucial testing ground for new technologies and approaches that would later be adopted by commercial broadcasters, and the community functioned as a testing ground for them.

Today, ham radio continues to be a popular activity for a large number of people all over the globe. It enables individuals to speak with others across huge distances while only requiring a few watts of power transmission. His legacy lives on in this dynamic community of amateur radio enthusiasts who continue to push the frontiers of what is possible with wireless communication technology. Although a lot has changed since Marconi's early experiments with wireless telegraphy, his legacy continues to live on in this community.

Beyond World War II: The Development of the World

The communications system was significantly impacted by the contributions of ham radio operators during World War II. During the war effort, a significant number of these amateur radio enthusiasts were recruited by various government agencies and military organizations to offer communication lines. Through the use of shortwave frequencies that were capable of traveling across extensive distances, these operators were able to interact with other hams located all over the globe. Following the end of the war, a significant number of these ham radio operators discovered that they had accumulated spare equipment and a newly discovered desire for communication. Through worldwide relationships, they established clubs and organizations with the goals of improving the pastime, facilitating the exchange of information and experiences, and fostering goodwill among countries to promote international cooperation. For the decades that have passed after World War II, HAM radio has continued to develop with the progression of technology. There are thousands of enthusiasts all over the globe who are part of the modern amateur radio community. These enthusiasts utilize a broad variety of radios, ranging from conventional analog radios to cutting-edge digital modes and satellite-based communications. Nevertheless, hams continue to have a profound regard for history and tradition, which is a way of paying tribute to those who came before them, despite all of these changes.

The Role of Ham Radio in Popular Culture: TV and Film

During the early 20th century, Ham radio was in existence. This activity requires the use of radios to speak with other ham radio operators located in different parts of the globe.

Throughout the years, this pastime has been portrayed in several films and television series, which is evidence of its widespread appeal and significant role in popular culture. Throughout the years, ham radio has been included in a variety of films and television series, which has contributed to its rise to prominence as a significant component of popular culture. As a result of its capacity to bring together people from all over the globe, it has become an intriguing activity for a great number of people, and the fact that it is now being portrayed on film serves as a reminder of the importance it has within our society.

Existing Situation: Is It a Decline or Resurgence?

The present situation of Ham radio is a subject that has been the subject of much discussion among both avid enthusiasts and influential figures in the industry. Others feel that there is a rebound in interest and engagement in the hobby, even though some people say that the hobby is seeing a decline. Many people, on the one hand, refer to the fact that the demography of ham radio operators is becoming older as proof of the problem. The number of younger persons who are willing to step up to take the position of elder members is decreasing as they retire or die away. As a further point of interest, developments in technology have made it simpler for individuals to interact with one another without the use of conventional radio apparatus.

On the other hand, many contend that young people are showing a fresh interest in ham radio and that newly developed technology is helping to feed this rebirth. Reddit and Discord are two examples of social media platforms that have enabled the creation of online communities that are oriented on the discussion and collaboration of ham radio. Furthermore, during times of natural catastrophes, when other modes of communication may be unavailable, emergency response groups sometimes depend on ham radios as a means of communication. While it is yet unclear what the future holds for Ham radio, it is quite evident that both supporters and opponents will continue to argue about whether or not we are now experiencing a decline or a comeback of the medium.

Possible Futures: What Does the Future Hold for Ham Radio?

The future of ham radio is bright, even if technological progress is still being made all around the globe. The incorporation of digital modes into ham radio communication is one of the developments that are anticipated to take place. There are more complex capabilities available with digital modes, such as encryption and error correction, in addition to quicker and more reliable transmission with digital modes. This may appeal to a new generation of enthusiasts who place a high importance on communication via high-tech means. The possible role that ham radio might play in emergency response scenarios is yet another future option for the technology. Ham radio operators have been renowned for a long time to offer essential communication during natural disasters; but, with developments in technology and training, their position may become even more important in the future. There is a possibility that governments and organizations may begin to allocate more resources toward the construction of powerful ham radio networks to help disaster relief operations.

Last but not least, there is the possibility that ham radio communication may place a greater emphasis on satellite technology. Amateur satellite projects have been gaining appeal among enthusiasts as tiny spacecraft become more widely available and inexpensive. Rather than depending on conventional infrastructure such as cellular towers or internet cables, these satellites may be used to promote communication on a worldwide scale. In general, the possibilities for the future of ham radio are quite exciting, and they are full of opportunities for the community to innovate and expand. Since it can be used for a wide variety of purposes, including but not limited to emergency communication, satellite tracking, and weather monitoring, ham radio has a tremendous amount of potential. In general, the history and promise of ham radio cannot be denied on any level. It is a dependable method of communication during times of disaster, and it also offers enthusiasts from all over the globe a delightful activity to participate in. Within the context of our ever-evolving world, it will be fascinating to see how ham radio adjusts to the ever-changing technological landscape and continues to fulfill its intended function.

Significance and Role in Modern Communication

Ham radio is a mode of communication that is based on the use of radio frequencies for purposes that are not related to commercial activity. In addition to having a long and illustrious history, it continues to be an important component of contemporary communication. The following are a few of the most important factors that emphasize the significance of ham radio in the modern world:

Emergency Communication

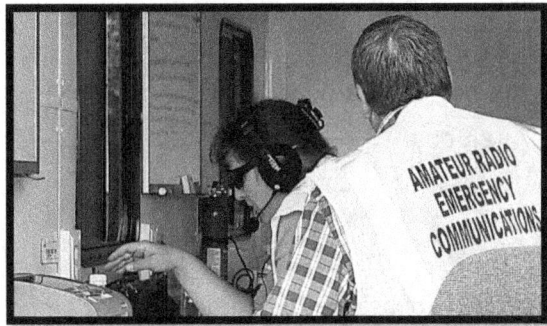

- **Disaster Response**: Ham radio operators are often the first to offer communication in the event of natural disasters or other emergencies, particularly in situations when established communication infrastructures are unable to function properly. They can provide vital information to emergency services, which helps in the efforts to respond to disasters.

- **Community Resilience**: Since Ham radio networks are decentralized and can function independently of centralized communication systems, they are very useful in circumstances in which the regular infrastructure is subject to disruption.

Global Connectivity

- **International Communication**: Ham radio makes it possible for people all over the world to speak with one another, which ultimately leads to the development of international friendships and relationships. The significance of this cannot be overstated, especially during times of geopolitical turmoil, when conventional avenues of communication may be blocked.
- **Ham Radio Satellites**: Ham radio operators contribute to the operation and maintenance of amateur radio satellites, which provide a method of communication in regions with inadequate terrestrial infrastructure. Ham radio satellites are a type of amateur radio equipment.

Technological Innovation

- **Experimentation and Innovation**: The community of amateur radio operators has a long history of making contributions to the improvement of technology via the use of experimentation and innovation. Hams often engage in the process of experimenting with and developing new technologies, expanding the bounds of radio communication, and making a contribution to the advancement of technology in general.
- **Education and Skill Development**: Learning about electronics, radio frequency engineering, and other technical skills may be accomplished via the use of ham radio, which offers a platform for people to learn about these topics. An interest in STEM (science, technology, engineering, and mathematics) subjects is cultivated by the implementation of this educational component.

Community Service

- **Public Service Events**: Volunteering their services for public events like marathons, parades, and festivals is something that Ham radio operators do as part of their participation in public service activities. They provide help for communication to guarantee the safety of these activities and to ensure that they are coordinated.

- **Search and Rescue**: Ham radio operators may be of assistance in search and rescue efforts by building communication networks that help in the coordination of rescue teams. This can be done in isolated places or in times of emergencies.

Personal Development and Social Connection

- **Personal Enrichment**: Ham radio is a hobby that enables people to continuously acquire new skills and develop new abilities, which also contributes to personal enrichment. It promotes a feeling of personal success and has the potential to be a pursuit that lasts a lifetime.
- **Community Building**: Local amateur radio clubs and organizations foster a feeling of community among enthusiasts, which is one of the potential benefits of community building. The sharing of information, experiences, and skills is further encouraged as a result of this.

Distinguishing Features of Ham Radios

There are several ways in which Ham radios are different from commercial radios. They provide characteristics that are one of a kind and are designed to meet the varied requirements of the amateur radio community.

The following is a list of features that characterize ham radios:

1. **Frequency Bands**

Ham radios can function on certain frequency bands that have been designated by regulatory bodies, especially for the use of amateur radio. Due to the frequency variation in these bands, operators can communicate on a local, regional, or even global scale.

Band	Low MHz	Low λ	Mid MHz	Mid λ	High MHz	High λ
2200	0.136	2205.9	0.137	2189.8	0.138	2173.9
630	0.472	635.6	0.476	630.9	0.479	626.3
160	1.800	166.7	1.900	157.9	2.000	150.0
80	3.500	85.7	3.750	80.0	4.000	75.0
60	5.352	56.1	5.360	56.0	5.367	55.9
40	7.000	42.9	7.150	42.0	7.300	41.1
30	10.100	29.7	10.125	29.6	10.150	29.6
20	14.000	21.4	14.175	21.2	14.350	20.9
17	18.068	16.6	18.118	16.6	18.168	16.5
15	21.000	14.3	21.225	14.1	21.450	14.0
12	24.890	12.1	24.940	12.0	24.990	12.0
10	28.000	10.7	28.850	10.4	29.700	10.1
6	50.000	6.0	52.000	5.8	54.000	5.6
2	144.000	2.1	146.000	2.1	148.000	2.0
0.7	420.000	0.7	435.000	0.7	450.000	0.7

2. **License Requirement**

The use of a ham radio, in contrast to the use of many other communication devices, requires the acquisition of an amateur radio license. It is the responsibility of the license holder to demonstrate that they have a fundamental comprehension of radio operation, rules, and technical issues.

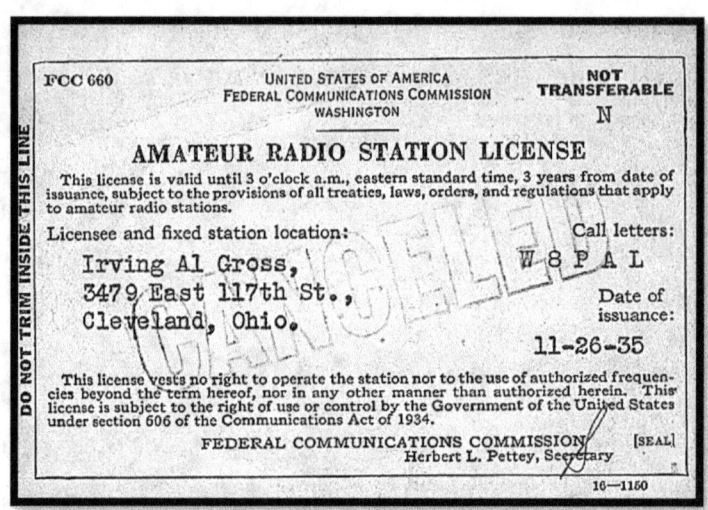

(An example of a Ham Radio license)

3. **Diversity of Modes**
- **Single Sideband (SSB), Frequency Modulation (FM),** and **Amplitude Modulation (AM)** are the three modes of communication that are supported by Ham radios. Another method of communication that Ham radios offer is voice. Because of this flexibility, operators can choose the mode that is most suitable for meeting their communication requirements.
- **Digital Modes:** Ham radios enable digital radio communication modes such as Morse code, packet radio, and others in addition to classic voice radios. More effective and error-free communication is made possible by digital modes.

4. **Power Output**

Ham radios often come with power levels that can be adjusted, which enables operators to choose the output power that is most suitable for their particular communication needs. For effective power management and the ability to respond to a variety of propagation circumstances, this functionality is very necessary.

5. **Ability to Communicate in an Emergency Situation**

An emphasis on public service is placed on the fact that Ham radios are an essential component of emergency communication. There are a lot of operators who are active in public service activities, such as providing communication assistance during different kinds of emergencies, catastrophes, and community events. Since Ham radios are intended to run on battery power, they are appropriate for use in emergency circumstances when regular power sources may be unavailable. This makes them acceptable for use in situations where standard power sources are unavailable.

6. **DIY projects and DIY experiments**

The amateur radio community has a long-standing legacy of kit-building, experimentation, and homebrewing techniques. A great number of operators take pleasure in building their equipment or altering existing radios to fulfill their particular requirements.

7. **Global Connectivity**

Ham radios make it possible to communicate across large distances, making this kind of communication known as DXing (distance communication). Many operators make it a point to establish connections with other stations located all over the globe to cultivate a feeling of community on a global scale.

8. **Spectrum of Activities**

The amateur radio community routinely conducts competitions and events that test the operators' abilities in a variety of radio communication-related domains. These contests and events encourage friendly rivalry and a sense of camaraderie among the participants. Ham radios can also be used for communication with amateur radio satellites that are circling the Earth. This enables operators to establish connections using space-borne repeaters.

9. **Frequency Agility**

Ham radios can span a broad frequency range, which enables operators to communicate on a large number of different bands. This versatility is significant since it allows for circumstances of propagation to change and different communication requirements to be met.

10. **Community and Social Aspects**

Amateur radio operators often take part in local clubs and networks (nets) to develop social relationships, exchange information, and provide support to one another. The whole experience of being an amateur radio operator is improved significantly by the presence of this community component.

Licensing and Regulations

FCC Licensing Requirements in 2024

The Federal Communications Commission (FCC) is in charge of administering and licensing the electromagnetic spectrum for both commercial and non-commercial uses, which includes state, county, and municipal governments. Public safety, commercial and non-commercial fixed and mobile wireless services, broadcast television and radio, satellite, and other services are all included in this category. Through the process of licensing the spectrum, the Commission advocates for efficient and dependable access to the spectrum for a wide range of creative applications, as well as for the promotion of public safety and emergency response actions. The Communications Act of 1934 is the law that governs the Federal Communications Commission (FCC), which is responsible for regulating amateur radio. In addition to this, it is bound by a great number of international accords. A license is required for all amateur radio operators. In the United States, there are three subcategories of licenses. As the class of license increases, the number of frequencies that are accessible increases as well. This means that to get each higher-class license, you will need to pass a more challenging test. Exams for licenses are administered by volunteer organizations of Amateur Radio operators, even though the FCC is in charge of such matters.

Under the supervision of groups known as Volunteer Examiner Coordinators, volunteers are responsible for the administration and grading of exams, as well as the reporting of the findings to the Federal Communications Commission (FCC), which is subsequently responsible for issuing licenses. United States licenses are valid for ten years before they need to be renewed, and anybody may possess one, except for a representative of a foreign government.

License Restructuring

The Federal Communications Commission (FCC) started making substantial revisions in December 1999, after a protracted examination of the Amateur Radio licensing structure. As of April 2000, there are just three license classes available, which is a reduction from the previous six classes. In addition, the Federal Communications Commission (FCC) stopped mandating Morse code competency in February of 2007. These new rules were published by the Federal Communications Commission (FCC) to simplify the licensing process and bring the Amateur Radio service into the digital age. Even while the new license system may not make it simpler to enter the world of amateur radio, licensed operators can advance more rapidly from the novice level to the expert level.

Technician License

The Technician class license is the most popular option for novice ham radio operators when it comes to obtaining their first license. Receiving a passing score on a single test consisting of thirty-five questions covering radio theory, rules, and operating procedures is necessary to get the Technician license. Those who have this license are granted access to all Amateur Radio frequencies that are higher than 30 megahertz, which enables them to speak with one another in their immediate vicinity and, more often, within North America. In addition to this, it grants access to some restricted privileges on the high-frequency (HF) bands, which are also referred to as **"short wave"** channels, which are used for international communications.

General License

Additionally, the General class license provides the holder with certain operating rights across all Amateur Radio channels and modes of operation. Because of this license, conversations may now take place all over the globe. A test consisting of thirty-five questions must be passed to get a license for the General class. Those who have licenses in the General class are required to have additionally completed the Technician written test.

Amateur Extra License

The Amateur Extra class license allows the holder to access all U.S. Amateur radio operators and is granted permission to operate on all bands and modes. The process of obtaining the license is more challenging; it needs passing a comprehensive test consisting of fifty questions. In addition, licensees of the extra class are required to have completed all written tests for the preceding license classes.

FCC Licensing Requirements for Ham Radios

1. **License Classes**

In the United States, the Federal Communications Commission (FCC) is responsible for issuing three different categories of amateur radio licenses: Technician, General, and Amateur Extra. The operational rights that are granted by each class are distinct.

2. **License Exams**

To receive a license to operate an amateur radio station, persons are required to pass a test that is given by Volunteer Examiner (VE) teams. These test candidates' knowledge of amateur radio in the areas of technical, regulatory, and operational aspects.

3. **Technical Class License**

The license associated with the Technician class is the entry-level license. Certain amateur radio frequencies are granted access to the benefits it offers.

4. **General Class License**

Compared to the Technician class, the General class license provides greater frequency rights on a wider variety of amateur radio bands. This is in contrast to the Technician card.

5. **Amateur Extra Class License**

The Amateur Extra class license is the highest level of the license system and provides the broadest operating rights across all amateur radio frequencies.

6. **Exam Elements**

Questions with multiple-choice answers are generally included in the exams. These questions cover a wide range of subjects, including regulations, operating procedures, radio wave propagation, and fundamental electronics.

7. **License Renewal**

The validity duration for an amateur radio license is 10 years. Before their licenses expire, licensees are required to renew their licenses.

8. **FCC Part 97 Rules**

Part 97 of the FCC regulations contains the rules that regulate amateur radio activities of the Federal Communications Commission (FCC). For amateur radio operators, it is very necessary to be conversant with these regulations and to adhere to them appropriately.

9. **Call Signs**

Individuals who have completed the examination are awarded a one-of-a-kind call sign by the Federal Communications Commission (FCC). It is possible to identify the individual's station on the air by using this call sign.

10. **License Upgrades**

Those who have licenses have the opportunity to update their licenses by completing higher-level tests, which grants them access to a greater number of frequency bands and modes.

11. **License Testing Locations**

Volunteer Examiner (VE) teams, who are often connected with amateur radio groups, are the ones who are responsible for conducting testing sessions. Exam sessions are conducted at a large number of different occasions and venues.

International Licensing Considerations

For communication, Ham radio operators make use of certain frequency bands. Every nation has its own set of national telecommunications agencies that are responsible for regulating the licensing of ham radios. The majority of the time, these agencies has their very own rules, regulations, processes, and procedures for licensing.

(ITU) International Telecommunication Union

This institution, known as the International Telecommunication Union (ITU), plays a significant role in the regulation of worldwide telecommunications, which includes the activities of ham radio. The International Telecommunication Union (ITU) is responsible for allocating frequency bands and providing guidelines; nevertheless, individual countries are in charge of implementing and managing their licensing procedures.

Important Considerations Regarding International Licensing

ITU Regions

There are three zones of the International Telecommunication Union (ITU), and each of these regions has its frequency allocations and regulations. Operators need to have a thorough understanding of the jurisdiction-specific regulations that pertain to them.

Reciprocal Agreements

Certain countries have reciprocal agreements that enable licensed ham radio operators from one nation to operate in another nation without the need to get a second license under the reciprocal agreement. For operators that want to engage with customers in other countries, it is essential to have a solid understanding of these agreements.

Licensing Classes

It is important to note that different countries have different licensing classes, each of which has its own set of rights and obligations. International operators must have a solid understanding of the equivalencies that exist between the various classes across the various countries.

Frequency Allocations

The International Telecommunication Union (ITU) designates certain frequency bands for use by amateur radio operators. On the other hand, specific countries may impose extra limits or allowances within these particular bands. Operators need to be conversant with the frequencies that are approved in the locations in which they want to communicate.

License Recognition

Foreign licenses may be recognized in some countries, while other countries need visitors to get temporary or special authorization to enter the country. When it comes to the recognition of licenses, it is very necessary to determine whether each nation has its special standards.

Call Sign Prefixes

The call sign prefixes of various countries are distinct from one another. It is essential to have a thorough understanding of the prefixes that are linked with each nation to successfully identify operators and guarantee compliance with licensing regulations.

Regulatory Changes

Over time, regulatory frameworks for ham radio will change. Operators need to be up-to-date on any new or updated licensing requirements and regulations that may be implemented in the countries in which they do business.

Tips for International Ham Radio Operations

They include the following:

- Before beginning operations in a new nation, do extensive study and make sure you have a solid understanding of the licensing requirements and regulations that are particular to that country.
- If you want the most up-to-date and correct information, you should get in touch with the appropriate regulatory authorities in the country in which you want to do business.
- It is recommended that you preserve copies of your license, any permits that are pertinent to your business, and documentation of your communication activities while you are participating in foreign business.
- Make it a habit to check for updates from the International Telecommunication Union (ITU) and national telecommunication authorities regularly to be updated about any changes in regulations.
- It is important to act following the guidelines established by the host nation and to be aware of the local traditions and regulations.

Staying Compliant with Local Regulations

For ham radio operators to guarantee that their equipment is used in a manner that is both safe and effective, they need to remain in compliance with local regulations. The pastime of ham radio, commonly referred to as amateur radio, is characterized by the use of specific radio frequencies that are assigned by regulatory bodies to maintain communication. By adhering to the precise regulations and guidelines that have been established by their respective local regulatory organizations, operators may ensure that they remain compliant.

Here's how to stay compliant with local regulations:

1. **Obtaining a license**

The acquisition of a ham radio license is often seen as a legal necessity in many countries. To do this, you will need to show that you have an understanding of radio regulations, operating procedures, and technical elements by passing an examination. There may be a variety of license classes in various countries, each of which confers a unique set of benefits.

2. **Frequency Allocation**

It is important to get familiar with the frequency bands that are designated for usage by amateur radio operators. There are a variety of reasons and traits that distinguish certain bands. Make certain that the bands that are specified for your license class are present in the operation of your equipment.

3. **Power Limits**

To comply with power restrictions, it is important to note that local regulations often stipulate maximum power levels for certain frequency bands. Exceeding these restrictions is not only against the rules but also has the potential to cause interference. You must check that the power level of your transmitter is adjusted appropriately.

4. **Station Identification**

Identifying your station following the regulations is also part of the proper call sign use. In most circumstances, this entails making a consistent announcement of your call sign while transmissions are taking place. Be sure that your call sign is displayed appropriately on all of your equipment.

5. **Station Location**

The area where you are allowed to operate your ham radio station may be restricted by certain regulations. Be aware of any geographical limits, and conduct your business solely within the regions that have been approved.

6. **Interference and a Monitoring System**

It is the responsibility of Ham radio operators to minimize interference with other radio services. These operators are responsible for avoiding interference. Your transmissions should be monitored to ensure that they do not interfere with other communications and any interference concerns should be addressed as soon as possible.

7. **Equipment Standard**

Make sure that your radio equipment is following the applicable technical requirements and certifications. When equipment is used in a manner that is not allowed or changed, there may be legal repercussions.

8. **Emergency Communication**

Emergency communication drills are something that many countries urge ham radio operators to take part in. Learn the protocols for dealing with emergencies and be prepared to assist if it is required.

9. **Privacy and Encryption**

Ham radio transmissions normally need to be open and accessible, thus you should avoid using encryption. The use of encryption or the concealment of the contents of your transmissions may be a violation of regulations.

10. **Renewal and Updates**

You are responsible for ensuring that your license is up to date and that it is renewed under the regulations that are in place in your area. Maintain a level of awareness of any modifications to the rules, and ensure that your understanding is up to date.

11. **Participation in the Community**

Participating in local amateur radio clubs or online forums may assist you in remaining up-to-date on the latest developments in the area, facilitating the exchange of experiences, and providing direction on matters about compliance.

Types of Ham Radios

Ham radios have several types of communication equipment that are used by amateur radio operators for non-commercial reasons. These uses include personal communication, communication in the event of an emergency, and investigation. There are many different sorts of these radios, and each one caters to a different set of requirements. A further categorization is based on mobility and use, such as mobile, base, and portable radios. The primary categories include transceivers, receivers, and transmitters. Other classifications are dependent on where the radio is used.

- **Transceivers**

Transceivers, which are all-in-one devices that combine the operations of sending and receiving signals, are referred to as transmitter-receivers, which is an abbreviation for the definition of the term. Due to their versatility, they are the most popular form of ham radio. They enable operators to converse on a broad variety of frequencies, making them the most frequent type.

- **Receivers**

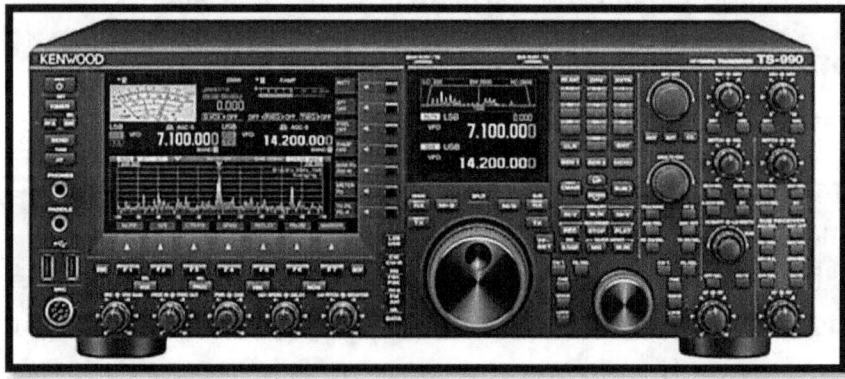

A receiver is a device that is specifically constructed to receive radio transmissions. Their inability to transfer signals is a significant limitation. Receivers are used by a significant number of ham radio operators to monitor various frequencies, explore the radio spectrum, and listen to broadcasts. Some receivers can be a component of a bigger station configuration that makes use of a separate transmitter.

- **Transmitters**

To put it another way, transmitters are equipment that are specifically designed to send out radio signals. Their inability to receive signals is a significant limitation. When it comes to ham radio stations, transmitters, and receivers are often associated with one another to establish a comprehensive communication system.

- **Mobile Radios**

This is a description of mobile ham radios, which are meant to be installed in vehicles such as automobiles, trucks, or boats. They have characteristics that are suited for mobile usage, and they are both small and versatile. In comparison to handheld radios, mobile radios often have a larger power output, which allows for a greater range of communication.

- **Base Radios**

Base station radios are used for permanent installations, which are commonly found in residential or commercial establishments. In comparison to mobile or portable radios, they offer a greater number of functions, have a higher power output, and have bigger antennas. For operators who have a specific site set aside for ham radio operations, base stations are the perfect solution.

- **Handheld Radios (HTs)**

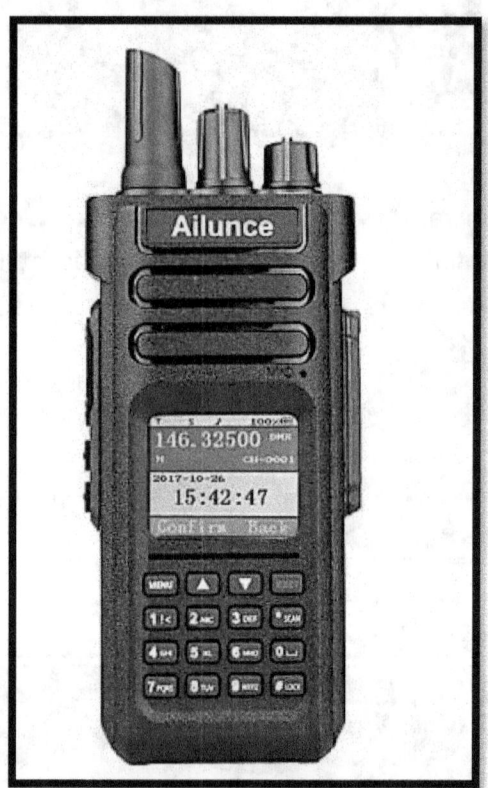

Handheld radios commonly referred to as HTs or portable transceivers, are devices that are relatively small, powered by batteries, and intended for use in portable settings. It is common for hams who need mobility or who are looking for an alternative mode of communication to use them. Since handheld radios often have a lesser power output in comparison to mobile or base radios, the communication range of handheld radios is typically shorter.

- **Dual-Band Radios**

Certain ham radios have the capability of operating on two distinct frequency bands, as described in the following sentence. Since they can move between bands depending on the circumstances or to meet certain frequency requirements, dual-band radios are very popular among operators who are looking for variety in their communication options.

- **All-Mode Radios**

All-mode radios are capable of functioning in a variety of modulation modes, including amplitude modulation (AM), frequency modulation (FM), and single-sideband (SSB). Following the communication needs and the radio frequency band that is being used, these radios provide the freedom to choose the modulation type based on the requirements.

The ham radio spectrum

With the use of transmitters, receivers, and antennas, ham radio, like other wireless technologies, makes use of the strength of electromagnetic radiation to communicate digital data, Morse code, and voices all over the globe. A sinusoidal wave is the manifestation of this electromagnetic energy as it moves across space.

The precise electromagnetic signal that you are dealing with is determined by the wave's wavelength as well as its frequency. A spectrum is a representation of electromagnetic radiation that can be broken down. It is divided into radio waves, microwaves, infrared, visible light, ultraviolet, x-rays, and gamma rays in the sequence of decreasing wavelength and increasing frequency.

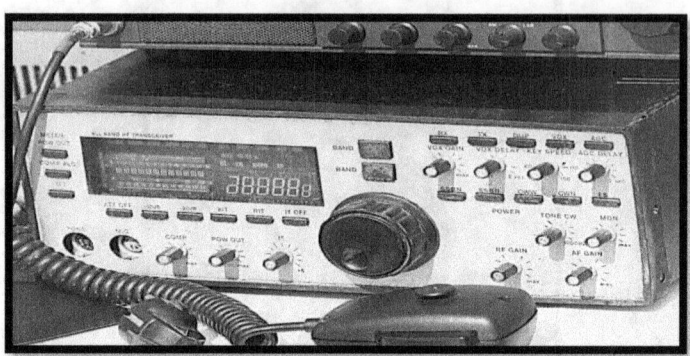

The radio wave spectrum is the only one in which ham radio functions. It is well known that the wavelengths of the radio wave spectrum are quite long, and they may span anywhere from 0.04 inches to over 62 miles! However, the specifics go into much more depth. Once this process is complete, the radio frequencies are separated into another spectrum, which is referred to as the radio frequency spectrum. To reserve certain bands of frequencies for particular radio technology, the Federal Communications Commission (FCC) has divided up this spectrum. Maritime radio communications, for instance, are carried out in the Very Low Frequency (VLF) band, while satellite communications are carried out in the Extremely High Frequency (EHF) band. Regarding ham radio, the Federal Communications Commission (FCC) has allotted a certain range of frequencies that begin at the AM radio band at 1.6 MHz and go all the way up to 1240 MHz. Very High Frequency (VHF) and Ultra High Frequency (UHF) are the two radio frequency bands that are included in this range. Every one of them has both advantages and disadvantages.

Very High Frequency (VHF)

The high-frequency (VHF) band is located on the radio frequency spectrum between 30 and 300 MHz, with the ham radio band being allocated for 144-148 MHz within this range. An example of a simplex communications system is provided by VHF. Between two ham radios, it enables line-of-sight communication. This band has a high degree of dependability and is less vulnerable to noise from electrical equipment that is located nearby. Because of this, it is the band that the majority of ham radio operators like to use. Ham radio operators often make use of repeaters that have been installed all across the nation by local radio clubs while they are conversing in the VHF band. These massive constructions that resemble antennas can receive and re-broadcast signals that are sent from a ham radio, which considerably expands the reach of the radio. Even better, a good many of these repeaters are powered by solar energy or have a power backup system built right in. In times of emergencies, this makes them ideal for maintaining communication channels.

Ultra-High Frequency (UHF)

Moving up the radio frequency spectrum, there is the Ultra High Frequencies, which span from 300MHz to 3GHz. These frequencies are quite intense. It is recommended that ham radio operators utilize the frequency band that extends from 420 to 450 MHz. UHF radio waves have a far shorter wavelength than VHF radio waves, and they are susceptible to interference from almost any solid object. VHF radio waves are known for their dependability. Consider the case of a skyscraper that is obstructing your signal or even your body. One of the advantages of using UHF for communication is that it occupies a greater portion of the available bandwidth, and it also offers a larger frequency range and superior audio signal quality.

Understanding Frequency Bands

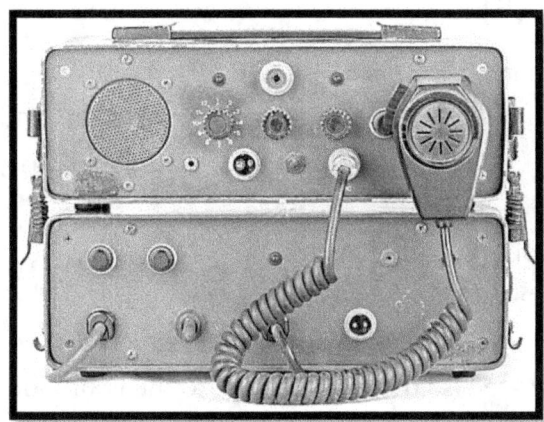

For amateur radio operators to efficiently communicate with one another and make the most of their radio equipment, they need to have a solid understanding of the frequency bands used in ham radio. Regulatory entities, such as the International Telecommunication Union (ITU), in conjunction with national telecommunications authorities, are responsible for determining the frequency bands that are allotted to ham radio operators.

Frequency Bands

Very High Frequency

- The range of frequencies is from 30 MHz to 300 MHz.
- Commonly used bands include 6 meters (50-54 MHz) and 2 meters (144-148 MHz).
- VHF signals can travel in a straight line and are appropriate for communication within a limited geographic area. Line-of-sight communication is a common use for them, and they can bounce off the ionosphere, allowing for greater ranges.

Ultra-High Frequency

- The frequency range is from 300 MHz to 3 GHz.
- The most often used bands are 70 centimeters (420-450 MHz).
- Ultra-high-frequency (UHF) transmissions are less susceptible to the effects of atmospheric conditions and can pass through obstructions more effectively than VHF signals. The fact that this is the case makes UHF an appropriate medium for local communication and operations in urban settings.

High Frequency (HF)

- The frequency range is from 3 MHz to 30 MHz.
- The following frequencies are the most often used bands: 80 meters (3.5-4.0 MHz), 40 meters (7.0-7.3 MHz), 20 meters (14.0-14.35 MHz), and 10 meters (28.0-29.7 MHz").
- The use of ionospheric propagation is one of the characteristics of high-frequency bands, which enables long-distance communication. The ionosphere acts as a reflector for signals, which enables communication on a worldwide scale. On the other hand, they are also more vulnerable to the conditions of the atmosphere.

Medium Frequency (MF) and Low Frequency (LF)

- The frequency range is as follows: MF—300 kHz to 3 MHz, and LF—30 kHz to 300 kHz.
- It is not typical practice in amateur radio to utilize the low-frequency (LF) band; nevertheless, the medium-frequency (MF) band might encompass the 160-meter band (1.8-2.0 MHz).
- These bands have good ground wave transmission, which makes them appropriate for communication in regional areas. On the other hand, because of the size of the antennas that are needed, they are employed less often.

Licensing and Regulations

- To demonstrate their familiarity with radio regulations and operating procedures, amateur radio operators are required to receive a license that is suitable from their respective national regulatory body.
- Various license classes can have access to a variety of frequency bands and transmission power levels.

Band Plans

- Various modes of communication, such as voice, Morse code, and digital modes, are all specified by band plans, which are guidelines that establish the frequencies that are included inside a band.
- Adhering to band plans guarantees an effective and well-organized use of the spectrum, minimizes interference, and encourages the implementation of best practices in operation.

Antennas and Propagation

- When it comes to successful communication, having a solid understanding of the propagation properties of various bands is very necessary. The propagation of an event is affected by a variety of factors, including the time of day, sunspot activity, and atmospheric conditions.
- It is of the utmost importance to choose the appropriate antenna for a certain band. The effectiveness of antennas is optimized to the greatest extent possible at certain frequencies.

Frequently Asked Questions

1. What is Ham Radio?
2. What is the history of Ham Radio?
3. What are the FCC Licensing Requirements in 2024?
4. What are the different features of Ham Radio?
5. What are the types of Ham Radios?

CHAPTER TWO
GETTING STARTED WITH HAM RADIOS

Overview

Chapter two further delves into the world of Ham Radio while discussing how to choose the right Ham radio, how to set up your Ham radio station, and the different operating modes of Ham Radio.

Choosing the Right Ham Radio

Assessing Personal Communication Needs

When it comes to ham radio, it is essential to examine one's particular communication requirements to maximize the selection of equipment, frequencies, and modes of communication. When conducting this evaluation, it is necessary to take into account a variety of aspects, including the objective, location, available resources, and individual preferences.

Identifying Purpose

- Identify the major objective of the communication that is used via ham radio. Do you intend to use it for activities related to hobbies, emergency preparation, public service, or any mix of these?
- Casual encounters, competitions, or experimenting with various modes and bands are all examples of activities that might be considered hobbyist pursuits.
- To be prepared for emergencies, it may be necessary to have dependable communication under difficult circumstances.

Recognizing the Different Frequency Bands

- The usage of ham radio is contingent upon the allocation of certain frequency bands. Analyze which bands are the most suitable for your communication requirements.
- Lower frequency bands, also known as HF bands, are appropriate for communication over long distances, while higher frequency bands, such as VHF and UHF, are ideal for communication within a local area.

Considering Location

- Conduct a thorough analysis of the geographical settings in which ham radio operations will be carried out. Depending on the context, urban and rural settings may provide distinct communication issues.
- Contrary to rural locations, which may have fewer impediments yet need long-distance communication capabilities, urban areas may have a higher level of interference and noise.

Assessing Antenna Options

- Antennas are an essential component in the communication process of ham radio. Take into consideration the regulations and the space that is available for placing antennas.
- Take into consideration the kind of antenna (dipole, vertical, or beam) that is appropriate for your operating circumstances and other requirements.

Power Requirements

- Analyze the power sources and requirements. Decide on whether you will run your business from a fixed location, rely on battery power, or use alternate alternative energy sources.
- It is possible for emergency communication plans to include backup power sources to guarantee communication in the event of power shutdowns.

Choosing Equipment

- Select a transceiver that takes into account the requirements of your communication system. The power output, frequency coverage, and modulation modes are all important considerations to take into account.
- The use of portable and mobile transceivers is appropriate for field activities, although base stations may provide a greater number of functions for home-based configurations.

Mode of Operation

- Determine the modes of communication that you wish to employ, such as voice (SSB, FM), Morse code (CW), digital modes (PSK31, FT8), or a mix of these modes.

- The bandwidth requirements for various modes of communication may vary, and one method may be more suitable for certain communication contexts than another.

Considering Future Expansion

- Determine if your requirements for communication could change with time. Think about purchasing equipment that can accommodate modifications and extensions in the future.

Education and Obtaining a License

- You are responsible for ensuring that you possess the appropriate expertise and licensing for the frequencies and modes that you want to operate
- To maintain compliance with regulations and keep up with technological changes, it is vital to engage in ongoing education.

Personal Preferences

- When purchasing equipment, it is important to take into account personal preferences, such as how easy it is to use, how the user interface is designed, and how ergonomically it is designed.

Features to Consider in Modern Ham Radios

Throughout their history, modern ham radios, which are sometimes referred to as amateur radios, have undergone tremendous development, including cutting-edge technology and feature that have the purpose of enhancing communication capabilities. If you are looking for a contemporary ham radio, there are a few crucial elements that you should take into consideration to ensure that the radio is suitable for your particular requirements. It is important to take into consideration the following features:

Frequency Range

- Ensure that the ham radio has wide frequency range coverage, including VHF (Very High Frequency) and UHF (Ultra High Frequency) bands. This particular feature is referred to as "**extended frequency coverage**." Some versions may also be able to cover high-frequency (HF) bands, which are used for communication over great distances.

Modes of Operation

- **Single Sideband (SSB):** When opposed to amplitude modulation (AM), single sideband (SSB) radio frequency allows for voice communication to be carried out with more efficiency.
- **Frequency modulation**, sometimes known as FM, is a common method of communication in the local area that provides high-quality audio.
- Support for **digital communication modes** such as D-STAR, DMR, or Fusion, which improves signal quality and data transfer, is referred to as "**digital modes**."

Output Power

- When it comes to communication across long distances, having a higher output power is helpful. It is important to take into consideration the power output capabilities of the radio, particularly if you want to work in difficult circumstances or isolated places.

Receiver Sensitivity

- It is essential for efficient communication to have a receiver of excellent quality that has a high level of sensitivity, particularly in situations when the signal strength is low.

Frequency Stability

- Your radio will remain on the intended frequency if you have a frequency control system that is both stable and precise. This will reduce the amount of drift and interference that occurs.

Built-In Antenna Tuner

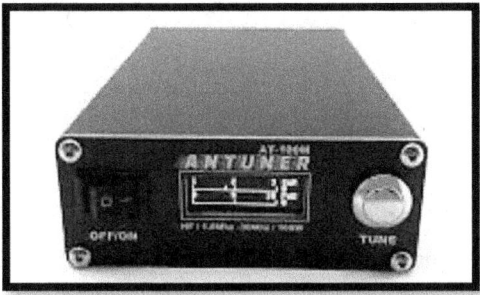

- An integrated antenna tuner can automatically match the impedance of the antenna system, which improves performance and decreases signal loss.

Digital Signal Processing (DSP)

- Digital signal processing (DSP) technology increases overall signal reception, lowers noise levels, and boosts audio quality. You should look for radios that have powerful digital signal processing characteristics.

Build-In GPS

- The capabilities of GPS enable automated location reporting and navigation functions to be implemented. When engaging in activities that need location monitoring, this might be useful for communication in the case of an emergency.

Multi-Function Display

- It is necessary to have a display that is both big and clear to provide simple navigation through menus, reading of signal information, and access to a variety of tasks.

Data Connectivity

- USB ports and data interfaces allow the radio to be connected to a computer, which enables the radio to get firmware updates, be programmed, and use digital data modes.

Remote Control Capability

- Some contemporary ham radios include remote control features that let you use a computer or mobile device to operate the radio from a distance.

Size and Portability

- Take into account the radio's dimensions and weight, particularly if you want to use it for mobile or portable activities.

Budget Considerations and Recommendations

A meticulous balance between price, features, and performance is required when it comes to budget considerations and suggestions for ham radios. A variety of equipment is used in the hobby of ham radio, commonly referred to as amateur radio, ranging from simple portable transceivers to complex base stations with cutting-edge features.

The following are some important financial factors and suggestions for ham radios:

Budget Considerations

- Before investing in radio equipment, take into account the price of acquiring an amateur radio license. Every nation has different licensing requirements, so be sure to account for these in your total spending plan.
- Decide whether you want to make a long-term investment or if you only need a basic setup to begin going. This will affect how much you are initially willing to invest.
- Various radios need distinct frequency bands to function. If you're interested in exploring certain bands, make sure the radio you choose is compatible with those frequencies. Greater cost may be associated with more costly radios that are more flexible and cover more frequencies.
- Choose between a fixed station arrangement and a portable handheld transceiver. Cost-effective portable solutions are often less expensive than permanent stations with greater power capacities.
- Enumerate the attributes that are necessary to meet your requirements. While basic radios are less expensive, be prepared to shell out more money for a more

feature-rich one if you need sophisticated functions like digital modes, APRS (Automatic Packet Reporting System), or satellite communication.

- In radio communication, antennas are essential. Set aside some cash for premium antennas and any other accessories you may need, such as connectors, coaxial cables, and antenna tuners.
- Take into account buying old equipment from reliable vendors. A lot of hams update their equipment, and reasonably priced, well-maintained radios are available. Make sure the secondhand equipment is in excellent operating order, however.
- Set aside money for a dependable power source if you're establishing a stationary station. Certain radios could need an external power supply; therefore getting a reliable power supply is essential for consistent functioning.
- Set aside money for resources, training materials, or classes that will improve your understanding of ham radio operations. Part of the activity is always learning new things.

Budget Recommendations

- **Baofeng UV-5R**

For novices with limited funds, the Baofeng UV-5R is a well-liked and reasonably-priced portable transceiver. It is adaptable for usage by beginners and covers a variety of bands.

- **Yaesu FT-60R**

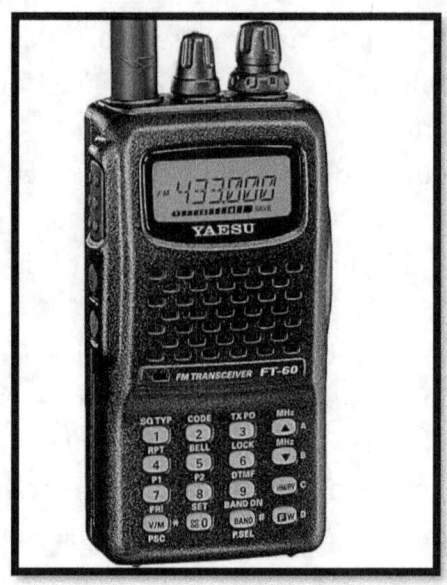

The Yaesu FT-60R is another portable option with a few extra functions; it's sturdy enough for outdoor usage and offers decent performance for its price.

- **Icom IC-7300**

It's well known that the Icom IC-7300 is a good mid-range HF transceiver. It has several features, including digital modes, a waterfall display, and an integrated tuner.

- **Kenwood TS-590SG**

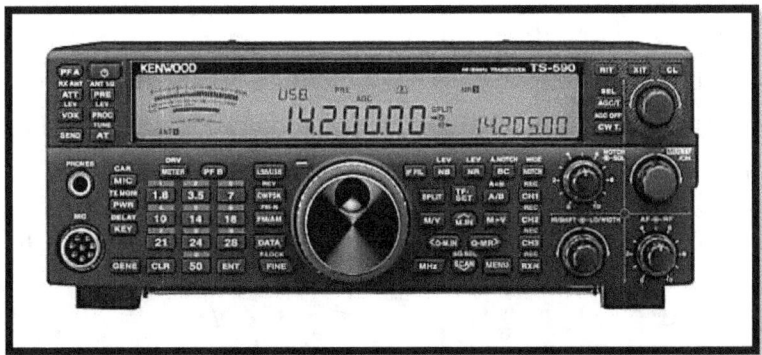

The Kenwood TS-590SG is a strong HF and 6-meter transceiver that offers a performance boost in terms of evaluations for its receiver.

- **FlexRadio Systems Flex-6400M**

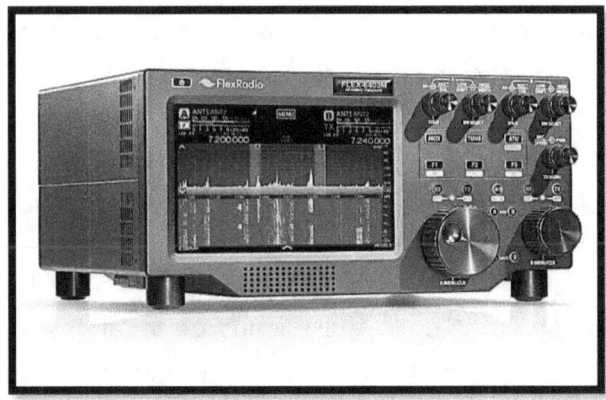

The Flex-6400M, an SDR (Software Defined Radio) with a modular design and extensive functionality, is a good option for individuals looking to spend more money.

- **DIY and Building Kits**

Take into account buying kits or constructing your equipment. This method may be economical as well as instructive, particularly for those who are curious about the technical parts of radio building.

The Role of Amateur Radio Operators in Emergency Communication

Natural catastrophes may happen suddenly, causing havoc and upsetting important communication systems. Amateur radio operators, commonly referred to as ham radio operators, are vital to improving disaster response operations during these difficult times.

Amateur radio operators play a crucial role in improving disaster response

These committed people provide a lifeline of communication that is sometimes the only dependable choice in emergency circumstances. They also bring equipment and knowledge to the table. When more conventional methods fail to create a link, amateur radio operators turn to wireless communication technologies. They are very significant assets during catastrophes because of their expertise and readiness, which guarantee that information can reach both emergency response professionals and impacted communities. Let's examine the benefits these operators provide and go further into the crucial role these operators play in disaster response.

Dependable routes of communication when all else fails

Communication networks may fail when a natural catastrophe occurs, isolating and stranding the impacted areas. Landlines, mobile networks, and the internet might all experience disruptions or outages. In these kinds of situations, ham radio operators intervene and use their equipment and knowledge to create trustworthy communication lines.

Benefits

- By using their own self-sufficient communication devices, such as antennas, backup power sources, and battery-operated radios, amateur radio operators may maintain contact even if public infrastructure fails.
- By using the High Frequency (HF) wavelength band, amateurs can communicate across longer distances than those possible with conventional communication methods. This guarantees communication between emergency response centers and impacted areas, regardless of their geographic separation.

- To increase coverage and optimize their reach during emergencies, amateur radio operators often organize networks and work together with other operators.

If regular communication channels are disrupted by a calamity, amateur radio operators may provide dependable and independent channels of communication. In times of emergency, their independence and capacity for long-distance communication are vital.

Rapid Deployment and Flexibility

Emergencies need quick reactions, and ham radio operators are renowned for their aptitude to quickly establish communication networks. With portable stations at their disposal, they may quickly set up makeshift radio stations in the impacted regions.

Benefits

- Most amateur radio operators belong to well-known emergency communication networks, including Radio Amateur Civil Emergency Service (RACES) or Amateur Radio Emergency Service (ARES). These organizations are effectively deployed during emergencies because they have established procedures and coordinated reactions.
- Operators can swiftly adjust to various frequencies and circumstances. To keep their devices operational, they are skilled in using a variety of bands and alternate power sources like solar panels or automobile batteries.
- Their equipment's adaptability enables them to establish communication from cars, shelters, or even mobile command centers, allowing them to coordinate with emergency response teams and provide critical information.

Amateur radio operators can quickly set up communication networks and guarantee timely information sharing in emergencies. They are very successful in coordinating disaster response activities because of their flexibility and agility.

Collaboration and Support from the Community

A common feature of amateur radio operators is their strong links to the community, which includes active participation in emergency management organizations. Their knowledge and selfless work are invaluable in fostering community unity, facilitating coordination efforts, and guaranteeing efficient catastrophe response.

Benefits

- Operators' train and exercise community members regularly to improve disaster readiness and make sure everyone know how to communicate if conventional modes of communication are unavailable.
- Amateur radio operators serve as an essential conduit between impacted areas and the wider response network by providing emergency management organizations with up-to-date information and important updates.
- Operators support emergency response teams during emergencies by relaying communications, giving situational awareness, and coordinating communication efforts between various organizations.

In addition to offering services beyond communication during emergencies, amateur radio operators are well-integrated within their communities. Their cooperation with neighborhood organizations guarantees efficient assistance and coordination, which eventually results in a more effective response to disasters. In emergency situations, amateur radio operators often go unnoticed as heroes. When all else fails, they can build dependable communication routes thanks to their commitment, knowledge, and willingness to assist.

They are crucial to improving disaster response operations because of their benefits, which include independent communication networks, quick deployment, flexibility, community cooperation, and assistance. It is impossible to overestimate the importance of these amateur radio operators as we continue to confront the growing difficulties posed by natural catastrophes. Their efforts eventually make a big impact in the face of hardship by helping to restore vital services, save lives, and offer crucial information.

How Amateur Radio Operators Maintain Connectivity during Emergencies: The Unbreakable Network

Using specific radio frequencies, amateur radio, often known as ham radio, is a leisure and service activity that enables people to connect. A robust network that can function even when all other modes of communication are down is provided by these radio enthusiasts, who are outfitted with their transceivers and antennas, which are essential to emergency response activities.

The Power of Amateur Radio Operators

Traditional communication infrastructure is susceptible to significant damage or disruption during natural catastrophes like hurricanes, earthquakes, or power outages. Amateur radio operators can help in this situation. They provide a lifeline that links those in need with essential resources by using their expertise, abilities, and access to technology.

In times of crisis, amateur radio operators are very useful due to their many important advantages:

- **Independence**: Amateur radio operators are self-sufficient, unlike traditional communication methods. They depend on their equipment, which is easily transported and swiftly assembled in any area. Due to their independence, they can provide communication solutions even in the most distant or remote locations.
- **Adaptability**: Amateur radio operators can adjust to changing conditions since they have access to a vast array of frequencies. When required, they might employ several modes of communication or switch frequencies to prevent interference.
- **Long-distance communication**: Amateur radio operators can create long-distance communication connections due to their sophisticated equipment and technological know-how. This skill is especially crucial in large-scale crises because the impacted areas could be dispersed or cut off from one another.
- **Robust network**: The huge network of amateur radio operators spans cities, regions, and even continents. Reliable communication is made possible by this

network, which may also readily grow to meet spikes in demand during emergencies.

Important Takeaways for Amateur Radio Operators

In an emergency or crisis, amateur radio operators may be of great assistance. A few important lessons learned are:

- **Community Support:** Amateur radio enthusiasts are often quite active in their neighborhoods, offering assistance during emergencies and communication services. To guarantee smooth coordination, they work together with law enforcement, emergency response teams, and other disaster relief groups.
- **Technical proficiency**: Through rigorous training and licensing procedures, amateur radio operators improve their technical proficiency. They have the know-how to maintain and operate a variety of radio equipment, which guarantees efficient communication under trying circumstances.
- **Emergency communication procedures**: By adhering to set emergency communication protocols, amateur radio operators guarantee effective and organized communication. They transmit vital information quickly and accurately, giving priority to emergency traffic.
- **Public service**: The hobby of amateur radio is focused on providing services to the public. Operators engage in training and drills to improve their abilities and believe it is their responsibility to help during emergencies. Their dedication to serving the public guarantees that they will always be prepared to respond in the event of a crisis.

The Impact of Amateur Radio Operators

It is impossible to exaggerate the importance of amateur radio operators in disaster preparedness and emergency response.

Let's examine some industry data that illustrates their significance:

- There are approximately 759,000 amateur radio licenses in the US alone, according to the American Radio Relay League (ARRL), making them a valuable resource during emergencies.

- When other systems collapsed during Hurricane Katrina, amateur radio was essential in establishing communication ties. The incident made their presence more crucial and raised awareness of and support for amateur radio emergency services.
- During Hurricanes Harvey, Irma, and Maria in 2017, amateur radio operators played a crucial role in emergency communication, easing coordination among emergency personnel and assisting impacted populations in remaining connected.

In times of crisis, amateur radio operators provide an impenetrable network that keeps communication going even when all other systems fall. They are significant assets in emergency response operations due to their independence, adaptability, and technical competence. Through the use of this robust and easily accessible communication method, communities may improve their capacity for disaster planning and response. Never forget that amateur radio operators are available to bridge the gap and connect those in need when all else fails.

Let us take a more in-depth look at some of the most prominent occasions in which Amateur Radio Operators played an important part in helping to resolve emergencies:

1. **(2005) Hurricane Katrina**

When Hurricane Katrina made landfall on the Gulf Coast of the United States, it inflicted enormous destruction and left millions of people without access to electricity or their ability to communicate. The establishment of communication channels and assistance with emergency response operations were both activities that were carried out by Amateur Radio Operators. They were responsible for communicating key information about those who had survived, coordinating the distribution of relief materials, and serving as an essential link to the outside world.

2. **Earthquake in Nepal in 2015**

As a result of the tragic earthquake that occurred in Nepal, the local telecommunications infrastructure sustained catastrophic damage. Participating in the rescue operations were amateur radio operators from all around the globe who joined together to provide their support. They offered services for emergency communication, participated in the search for those who had gone missing, and gave assistance to international organizations helping to coordinate relief.

3. **Hurricane Maria, which struck Puerto Rico in 2017**

Following the devastation caused by Hurricane Maria, Puerto Rico was left without electricity and communication over the entire island. Amateur Radio Operators swiftly mobilized, establishing temporary radio stations and providing emergency support. Their efforts included facilitating communication among emergency responders, assisting in the reunification of families, and contributing to the restoration of essential infrastructure. In times of emergencies, the role that amateur radio operators play in maintaining communication is crucial. Both their readiness and their dedication are shown by the fact that they can build communication channels if regular techniques are unsuccessful. If all else fails, these highly qualified operators are there to give a lifeline to those who are in need. Despite the quick pace at which technology is advancing, it is critical to recognize the everlasting significance of amateur radio in the event of an emergency. It is because of the commitment and competence of these operators that communities that are experiencing difficulties as a result of catastrophes can maintain their connections, get aid, and eventually rebuild. To facilitate efficient communication in times of emergency, amateur radio operators are mobilizing communities.

Setting Up Your Ham Radio Station

Establishing your radio station can be a satisfying hobby since it allows you to communicate with people all over the world, take part in efforts to enhance communication during times of emergency, and experiment with a variety of technology. What are the stages involved in establishing your radio station?

Understanding Amateur Radio Communications Concepts

Ham radio, also known as amateur radio, is the practice of using radio frequencies for non-commercial activities, such as conducting wireless experiments, sending messages, engaging in self-training, engaging in private enjoyment, participating in radio sports, competing, and communicating in an emergency situation. Not only is it a pleasant, sociable, and instructive activity, but it may also be quite helpful in times of emergencies. Anybody can begin participating in this intriguing sport if they have the necessary information and equipment.

Preparing for Your Amateur Radio Station Setup: Tools and Equipment You'll Need

To be ready for the establishment of an amateur radio station, you will need to have certain important tools and equipment.

Here are some of the fundamental things that you will require:

- **Amateur Radio License**: Before you set up your station, you need to get an amateur radio license from the telecoms authorities in your nation since it is a requirement.
- **Microphones**: One of the most fundamental pieces of equipment is the microphone.

They listen to the voice of the operator and assist in transmitting it to everyone who is listening.

- **Transmitter**: The most essential component of ham radio equipment is the transmitter, which is of the highest importance. The signal from the microphone is received by this control panel, which then converts the sound into sine waves that are present.
- **Antenna**: When it comes to transmitting a broadcast for other people to hear, the antenna is the most important component. A variety of sizes are available for the antenna that is attached to the amplifier. The ham radio operator will be prepared to be heard after they have acquired all of the necessary equipment, including an antenna.

- **Amplifier**: The next piece of ham radio equipment that is required for a broadcast is an amplifier. Ham radio amplifiers are essential for broadcasting. For the signal to be sent to listeners, it must first be amplified by an amplifier once it has been received.

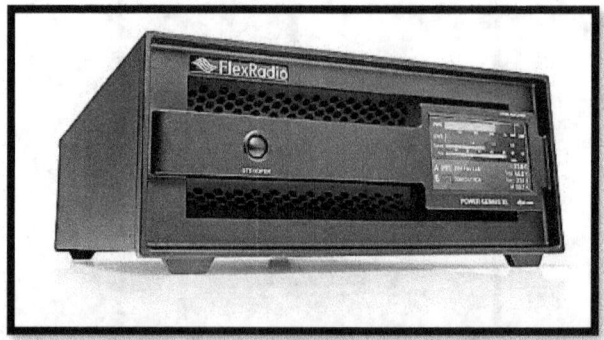

- **Coaxial Cable:** When you want to connect your radio to your antenna, you will need to use a kind of connection known as coaxial cable.

- **Grounding Equipment:** It is in your best interest to ground your equipment to prevent interference and ensure your safety. Installing grounding clamps, a grounding rod, and a grounding wire are all things that you will need to do.
- **Power Supply:** The power source for your radio might be a battery, a power supply unit, or even a generator. All of these options are viable options.

Building Your Antenna System: Types and Placement Considerations

There is a possibility that the performance and efficiency of the system may be significantly influenced by the kind of antenna and the location of the antenna.

The following are some important things to keep in mind while constructing your antenna system:

Several various kinds of antennas may be used, including Yagi antennas, patch antennas, dipole antennas, and pedestal antennas. Yagi antennas are the most common variety. The height of the antenna, any obstacles like buildings or trees, correct grounding to protect the antenna from lightning or increase signal quality, and the configuration of the antenna should be given the highest priority when it comes to the placement of the antenna. This may have a substantial effect on the system's performance as well as its dependability.

Finding the Transceiver That Is Best Suited to Your Requirements

Choosing a transceiver is dependent on several different aspects to take into consideration:

1. Determine the wavelength you want to use.
2. Confirm that you are compatible.
3. Consider the speeds.
4. Keep in mind the range.
5. Find out the recommended amount of power output.
6. Examine the world around you.

By taking into account these aspects, you will be able to choose the transceiver that is most suitable for your requirements and assure the highest possible performance in the application that you are working on.

Configure your transceiver

1. **Frequency and Mode Settings**: Adjust the frequency and mode settings so that they correspond to your preferences and the bands that you want to operate on.
2. **Power Output**: Make adjustments to the power output under the circumstances of the band and the requirements required for communication.

Installing and Configuring Your Radio Equipment

You must do so while installing and configuring your radio equipment.

1. Make sure that the place you choose is clear from any obstructions or interference.
2. In addition to mounting the antenna, check to see that it is grounded correctly.
3. Establish a connection with the power supply.
4. It is recommended to install supplementary components such as signal boosters and power amplifiers.
5. Following the instructions provided by the manufacturer, configure the settings on your radio equipment.

6. Check that the system is operating without any hiccups.

Acquire the Necessary Equipment

Basic Equipment

1. **Transceiver (Radio):** The radio transceiver that you choose should be capable of supporting the bands and modes that you are interested in. Think about things like the amount of power it produces, its size, and its features.
2. **Antenna:** Choose an antenna that is appropriate for the bands that you have selected. There are many different kinds of antennas, including vertical, dipole, and Yagi antennas, among others.
3. **Power source**: Make sure that you have a dependable power source that is compatible with the specifications of your transceiver.

Optional Equipment

1. **Tuner**: A tuner is a device that may assist in matching the impedance of your antenna to that of your transceiver, hence enhancing performance.
2. **Amplifier**: If you want to improve the quality of your signal, you should think about installing an amplifier to raise the amount of effective radiated power (ERP) that you have.
3. **SWR Meter**: A Standing Wave Ratio (SWR) meter is a device that assists you in adjusting your antenna so that it functions at its highest possible level.

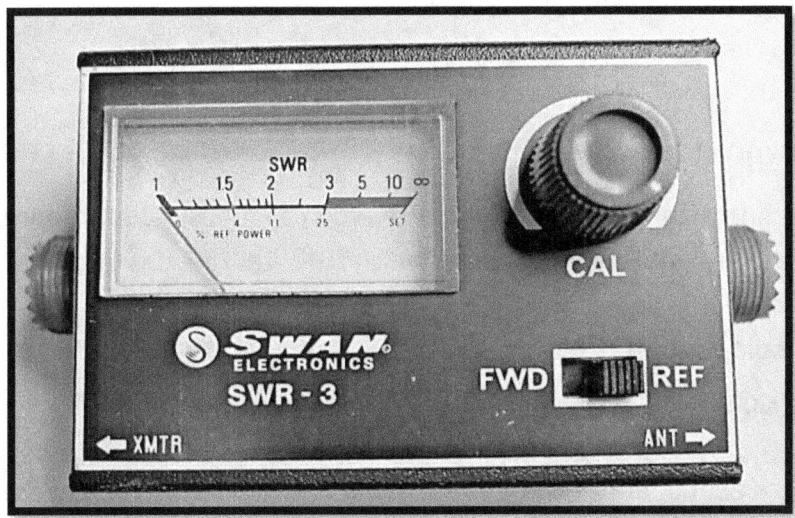

4. **Headphones and Microphone:** If you want to improve your experience with voice communication, making use of headphones and a microphone of high quality might be beneficial.

Obtaining Your Amateur Radio License and Call Sign

There are a few phases involved in the process of acquiring a call sign and license for amateur radio.

1. Get an FRN - To get an FRN, you will need to register with the FCC Cores first.
2. Search for a timetable and make arrangements with the VEs in your region to determine the test schedule.
3. Review - Begin your preparations for the forthcoming examination before you get the official schedule for the examination. There are three different levels of the license examination: the Technician level, which is the beginning level, the General level, and the Amateur Extra level, which is the highest level.
4. Pay the cost.
5. Take the exam.
6. Apply for a license using the Universal Licensing System (ULS) of the Federal Communications Commission (FCC) after you have mastered the test. You will be required to enter personal information as well as your Social Security number or Federal Register number, both of which may be obtained from the website of the Federal Communications Commission.
7. Select a call sign to use.
8. Get on the air.

Test and Adjust

1. **SWR Adjustment**: With an SWR meter, you can adjust your antenna such that it has the lowest SWR value possible, which indicates that it is performing at its best.
2. **Conduct Test Transmissions**: To verify that your apparatus is operating properly, you should conduct test transmissions.

Operating Modes

There are a variety of operating modes available for Ham radios, and each mode serves a distinct function. The following is a list of common modes of operation:

Single Sideband (SSB)

Single Sideband is a speech mode that transmits just one sideband of the audio signal, which allows for effective use of bandwidth. In the world of long-distance communication, notably in the HF (High-Frequency) bands, single sideband (SSB) is an extensively used method. The signal-to-noise ratio that it offers is superior to that of other speech modes, such as AM frequencies.

Frequency Modulation (FM)

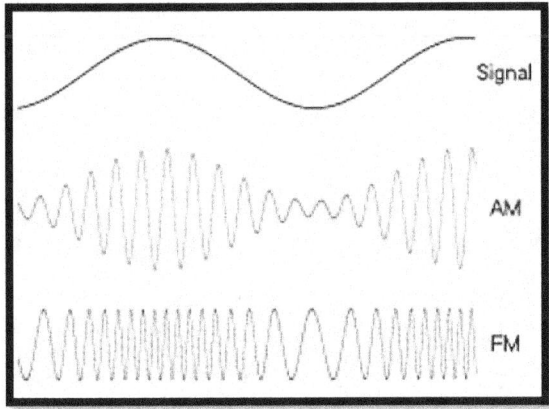

FM mode is used for voice communication and offers a satisfactory level of audio quality. It alters the frequency of the carrier signal in a modulated manner. The frequency modulation (FM) technique is often used for local communication in the very high frequency (VHF) and ultra-high frequency (UHF) bands. When it comes to repeater operations and short-range connections, it is used often.

Amplitude Modulation (AM)

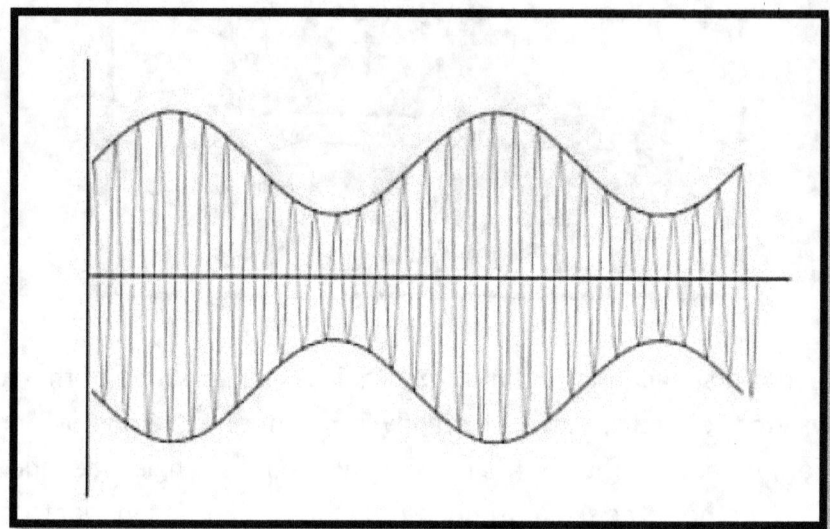

AM regulates the amplitude of the carrier signal, as described in the previous sentence. This mode is older than SSB or FM, and it has a greater bandwidth than any of those two. In addition to its application in broadcasting, AM is used in the high-frequency (HF) bands for voice communication. In contrast to SSB and FM, it is considered to be less frequent in amateur radio.

Digital Modes

The transmission of information is accomplished via the use of digital encoding in digital modalities. In addition to PSK31, FT8, and RTTY, these modes also contain others. Digital modes are useful in situations where there is weak propagation, and they can transfer text, photos, and other types of data. Competitions and low-power communication are two of the most common uses for them.

Continuous Wave Mode (CW)

CW is a way of communication that involves the transmission of information via the use of Morse code. This model is characterized by the on-off keying of the carrier signal. Broadcasting (CW) is used for long-distance communication, particularly in the high-frequency (HF) bands. When it comes to communicating with little power and weak signals, this method is an effective one.

Packet Radio

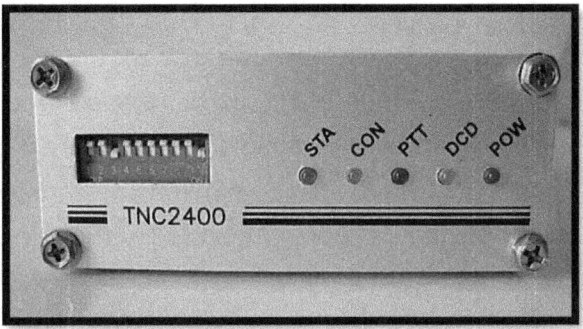

The transfer of digital data in the form of packets is part of the process of packet radio, which is analogous to the process of computer networking. The usage of packet radio includes the transmission of messages, the transfer of files, and other forms of data communication. Regularly, it is used in the VHF and UHF bands.

Satellite Modes

Ham radio operators can communicate with one another using satellites by employing certain modes ranging from FM to SSB to digital modes. The range and accessibility of communication are increased by the ability of ham radio operators to communicate with amateur radio satellites using satellite modes.

Repeater Operation

Repeaters are devices that receive signals on one frequency and concurrently retransmit them on another frequency. This allows them to expand the range of effective communication. Repeaters are often used in combination with FM modes, particularly in the VHF and UHF bands, to enlarge the geographical region that is covered.

Voice, Morse code, and Data Transmission

The exchange of voice signals across radio frequencies is known as voice transmission in signal radio, often known as amateur radio. For speech transmission, Hams makes use of a wide range of modulation methods, the most prevalent of which being Amplitude Modulation (AM) and Frequency Modulation (FM).

Amplitude Modulation (AM)

During AM transmission, the amplitude of the carrier wave is modulated by the audio signal, which is the voice. To facilitate communication across great distances, AM is often used in the HF (High Frequency) bands. In comparison to other modes, AM requires a greater bandwidth, which results in worse spectrum efficiency. Despite this, hams continue to make extensive use of AM.

Frequency Modulation (FM)

The frequency of the carrier wave is modulated by FM depending on the audio signal that is being received. Radiofrequency (FM) is often used for local communication in the VHF (Very High Frequency) and UHF (Ultra High Frequency) bands. Furthermore, in comparison to AM, it offers superior audio quality and is more resistant to interference that is dependent on amplitude.

Morse code in Ham Radio

The use of Morse code, which is a way of encoding text characters using sequences of dots and dashes, has a long and illustrious history in the field of ham radio and retains its widespread use for certain reasons.

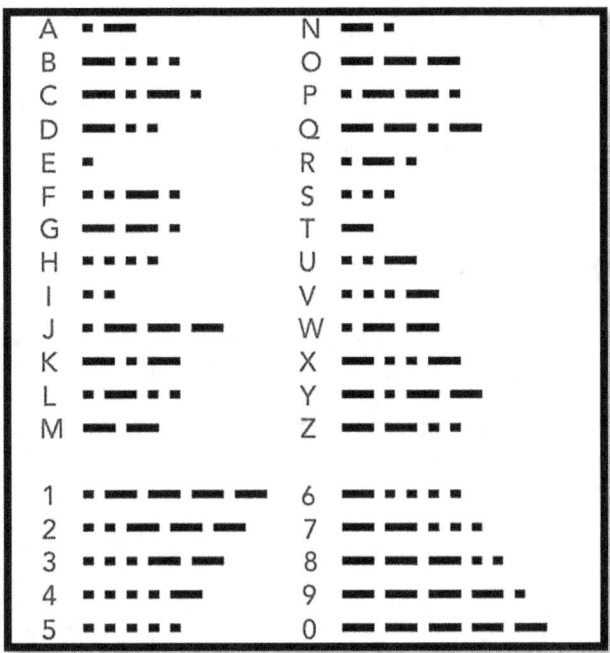

Morse code Components

- Individual letters or numbers are represented by dots, which are for short signals, and dashes, which are for lengthy signals.
- The timing of signals and the distance between them are very important for decoding.

Continuous Wave Mode (CW)

- The transmission of Morse code often involves the use of a carrier wave that is either active or inactive. Hams make use of a key to manually terminate the carrier wave, which results in the production of Morse code messages.

Use of Morse code in Ham Radio

- In case of an emergency, Morse code may prove to be more efficient than other ways of communication, particularly in situations when other mechanisms of communication are hampered. Since Morse code can be communicated at lower power levels, it is beneficial for communication that is more energy efficient.

Data Transmission in Ham Radio

Ham radio operators use a variety of digital modes for data transmission in addition to speech and Morse code. These modes allow for the transfer of text, photos, and other types of data.

Phase Shift Keying (PSK)

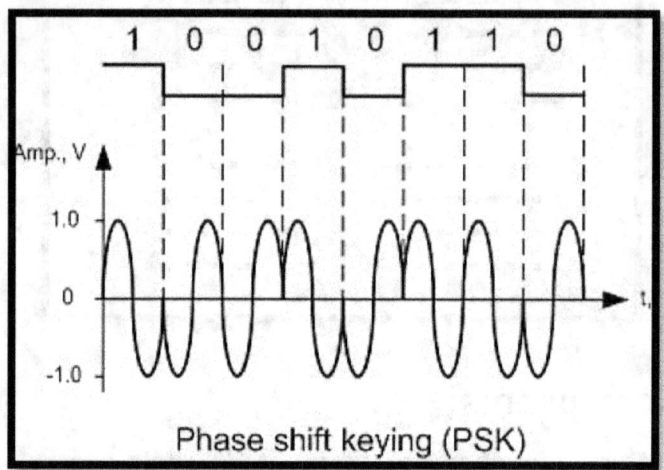

Phase-shift keying (PSK) changes the phase of the carrier wave to represent digital information. It is useful for keyboard-to-keyboard communication and is dependable even when subjected to difficult circumstances.

Radio Teletype (RTTY)

To represent binary digits, RTTY makes use of two distinct frequencies. It is an excellent choice for sending text messages over ham radio because of its versatility.

Packet Radio

For transmission, packet radio breaks data up into smaller units known as packets, which enables error detection and repair. The transmission of digital data and messages via ham radio networks is a common use of this technology.

Automatic Packet Reporting System (APRS)

This system integrates ham radio with global positioning system technology to provide real-time location reporting, weather data, and other information. It is used extensively to monitor mobile stations and for communication in times of emergency.

Frequently Asked Questions

1. How do you choose the right Ham Radio?
2. How do you set up your Ham Radio station?
3. What are the features to consider in modern Ham Radios?
4. How do you determine the budget considerations in Ham Radio?

CHAPTER THREE
LICENSING AND TECHNICAL BASICS

Overview

You cannot fully operate a Ham Radio or a Ham Radio station without obtaining a license. This chapter talks about the different licensing classes in Ham Radio and how you can obtain one.

Ham Radio Licensing Classes

Technician, General, and Extra Class Licenses

For persons to lawfully engage in activities related to ham radio, they are required to get a ham radio license. Aspiring ham radio operators might benefit from taking license courses, which are designed to assist them in preparing for and passing the necessary licensing exam countries. These classes include a wide variety of subjects, such as radio theory, regulations, and both theoretical and practical aspects of radio operation.

Classes for Ham Radio Licenses

Entry-Level Classes

Technician Class: The Technician Class is the licensing class that is considered to be entry-level. It discusses fundamental ideas in radio communication, as well as safety and the restrictions of the FCC. Technologists are allowed rights on several different frequency bands, which enable them to participate in communication on a local and regional level as necessary.

Intermediate Classes

General Class: Individuals who have obtained a Technician license are eligible to seek the General Class license. The topics of radio theory, propagation, and antenna design are explored in more depth in this course. Individuals who have General Class licenses are granted increased frequency rights, which enable them to communicate over greater distances and on a greater number of bands.

Advanced Classes

Extra Class: In the world of amateur radio licensing, the Extra Class license is considered to be the greatest degree of authorization. It requires a more in-depth grasp of sophisticated radio principles, as well as practical skills and laws applicable to the field. Extra-class operators can employ high-powered transmitters and have access to the most extensive spectrum of frequencies than other operators.

Key Components Covered in Ham Radio Licensing Classes

Radio Theory

- Understanding the electromagnetic spectrum, which includes the numerous frequency bands that are accessible to ham radio operators and how radio waves travel through the atmosphere.
- **Antenna Theory**: The fundamentals of antenna design, including the many kinds of antennas and the applications for which they are used.

Regulations and Standard Operating Procedures

- Familiarization with the regulations that regulate the operation of amateur radio stations is following the FCC regulations and Regulations.
- Etiquette, language, and processes that are appropriate for successful and polite communication are included in the field of operating practices.

Safety

- Handling and using radio equipment safely is classified as electrical safety.
- Protocols for ham radio operators to follow in the event of emergencies and disasters are referred to as emergency procedures.

Practical Skills

- The process of establishing a ham radio station, which includes the installation of antennas and transceivers, is referred to as "**station setup**."
- The On-Air Operating course consists of hands-on activities that allow participants to communicate with other operators and take part in ham radio events.

Teaching Methods

Classroom Instructions

- Experienced ham radio operators or teachers who provide in-person or online courses.
- Presentations, lectures, and interactive discussions to address theoretical elements of the topic.

Hands-on Training

- Hands-on training sessions on how to make radio equipment operational and set it up.
- Exercises that simulate being broadcast on the radio to build your communication abilities.

Online Resources

- Availability of online study resources, practice tests, and discussion forums for added convenience.
- In addition to the more conventional classroom education, there are also video tutorials and webinars.

Exams and Certification

Technician and General Exams

- Exams consist of multiple-choice questions that are given by teams of Volunteer Examiners (VE).
- To be eligible to take the General test, candidates must first pass the Technician exam.

Extra Class Exam

- An additional examination for those who are interested in obtaining the Extra Class license.
- Requires a greater degree of expertise in radio theory and regulations to be successful.

Exam Preparation and Resources

To be prepared for a ham radio test, it is necessary to have a solid grasp of the fundamentals of radio communication, as well as the associated rules and technical elements of amateur radio operation.

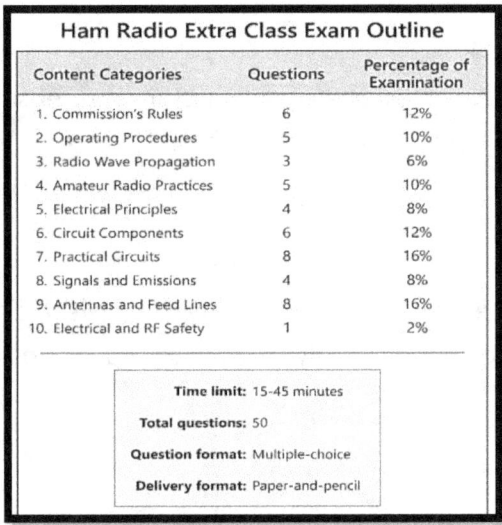

A comprehensive strategy for preparing for the ham radio test and gathering materials is as follows:

License Levels

- **Technician** provides privileges on both the VHF and UHF bands and is considered an entry-level license.
- **General**: It extends rights to include high-frequency bands.
- **Extra**: The highest level, with extra rights while using HF.

Exam Elements

- **Regulations**: FCC standards and operational procedures are included in the regulations.
- **Technical** includes fundamentals of electronics, antennas, and the propagation of radio waves.
- **Practical**: techniques for operating the station and setting it up.

Resources for Study

Books

- **ARRL licensing Manuals**: These manuals, which are published by the American Radio Relay League (ARRL), cover all of the different licensing levels.

Online Courses

- Courses are available for all licensing levels on the website HamRadioPrep.com.
- Interactive online courses offered by the ARRL are referred to as ARRL Online Courses.

Practice Exams

- There are practice exams available for all licensure levels on the website QRZ.com.
- HamStudy.org provides a study program that is both fully free and fully configurable.

Mobile Apps

- All licensing levels are covered by the Ham Test Prep app, which is available for both iOS and Android.
- HamRadioExam is yet another software that allows users to practice for tests while they are on the go.

Online Forums

- **QRZ Forums**: Have conversations with members of the ham radio community and ask for guidance.
- **eHam.net Forums:** An additional venue for conversations and questions with other users.

Practical Training

Local Clubs

- The ARRL Club Locator allows you to locate a local ham radio club that provides support and hands-on instruction.
- It is possible to search for local ham radio gatherings on Meetup.com.

FCC Regulations

FCC Part 97

- You should familiarize yourself with the laws that regulate amateur radio by reading them and understanding them correctly.
- Obtain access to the official papers and updates by visiting the FCC website.

Practice Tests

- In the ARRL Exam Review, you will go over genuine questions that were on prior tests.
- Additional resources for practice exams can be found at the W5YI Exam Review.

Additional Tips

- Make a schedule outlining particular times throughout the day to study.
- To make your study plan more achievable, you should establish objectives.
- To acquire practical skills, you can construct basic electronic kits.
- Participate in conversations, ask questions, and gain knowledge from other users on Reddit.
- Some servers are specifically designed for ham radio conversations.
- Keep up with the latest developments in technology and laws by following several industry news sources.
- If you want to get new perspectives, you should listen to ham radio podcasts and read blogs.
- Take pauses to prevent burnout and to keep your thinking from becoming stale.
- Listening to the ham radio should be a pleasurable experience rather than a stressful one.

Upgrading License Classifications

Amateur radio operators can obtain access to more frequencies, privileges, and forms of communication via the process of upgrading their license classes in amateur radio. The purpose of the ham radio licensing system is to guarantee that operators possess the knowledge and abilities required to operate radio equipment in a manner that is both safe and operationally efficient. The Federal Communications Commission (FCC) in the United States and Ofcom in the United Kingdom are two regulatory bodies that award ham radio licenses in numerous countries. **The following is an in-depth summary of the procedure that is followed to upgrade license classes in ham radio:**

Understanding License Classes

- **Technician Class**: Technician Class permits privileges on both the VHF and UHF bands. This is often the license that is considered to be the entry-level license.
- **General Class**: Provides extended rights on high-frequency bands, which made it possible to communicate across great distances.
- **Amateur Extra Class**: The highest level, which grants access to all amateur frequencies and modes.

Upgrade Requirements and Obligations

- **Knowledge and Study**: To upgrade, operators are required to study extra content that is particularly pertinent to the higher license class.
- Practical experience earned from working at the current licensing level is often necessary. This experience may be gained via experience.

Study Materials

- **License Manuals:** Official study guides that cover the subject for each class are provided by licensing authorities.
- **Online Resources**: There are a multitude of websites that provide applicants with practice tests, study guides, and other tools to assist them in preparing for the examination.

Taking the Exam

- **Scheduling**: Candidates are responsible for scheduling exams with groups of approved Volunteer Examiners (VE).
- **Exam Structure**: Exams normally consist of questions with multiple choice answers, and the number of questions and the minimum score required to pass vary depending on the kind of license being taken.

Exam Elements

- Questions about rules, regulations, and appropriate operating procedures are included in the "**Regulations and Operating Procedures**" category.

- Assessments on electronics, radio wave propagation, and station setup are included in the technical knowledge assessments.
- Demonstrating the ability to successfully operate radio equipment and execute fundamental calculations is included in the practical skills category.

Passing the Exam

- **Score Reporting:** The results of the examination are often supplied either immediately or soon after it has been finished.
- **The issuance of a new license**: Once the examination has been passed, the regulatory body will issue a new license that reflects the changed class.

Post-Upgrade Privileges

- **Expanded Frequency Bands**: The ability to communicate across greater distances by using extra high-frequency signals.
- Privileges to use sophisticated modes of communication, such as digital modes and satellite communication are included in the category of superior modes.

Community Involvement

- Joining amateur radio clubs and communities to establish professional connections and exchange information is an example of club participation.
- To create a learning atmosphere that is encouraging and friendly, experienced operators often act as mentors to beginners.

Basic Electronics for Ham Radio Operators

Understanding Resistors, Capacitors, and Inductors

Resistors

An electrical component called a resistor is used in circuits to restrict or regulate the passage of electric current. It is distinguished by its resistance, which is expressed in ohms (Ω). **Role in Ham Radio:**

1. **Current Limiting**: In a ham radio circuit, resistors are often employed to control the amount of current that passes through certain parts.
2. **Voltage Division**: They aid in the construction of voltage dividers, which are beneficial for biasing circuits or attenuating signals.

3. **Termination**: To prevent signal reflections, transmission cables are terminated using resistors.

Types:

1. **Fixed Resistors**: These resistors have a fixed resistance.
2. Potentiometers, or variable resistors, provide resistance that can be adjusted.

Based on the particular needs of your ham radio circuit, choose resistors with the right power ratings, tolerance levels, and resistance values.

Capacitors

An electrical component with two terminals that stores electrical energy in an electric field is called a capacitor. Its capacitance, expressed in farads (F), is what defines it.

Role in Ham Radio:

1. **Filtering**: To exclude undesired signals or noise, filters use capacitors.
2. **Coupling**: They obstruct DC components while facilitating the passage of AC signals.
3. **Energy Storage**: To help with peak power needs, capacitors store energy and release it as required.

Types

1. **Electrolytic Capacitors**: High-capacitance electrolytic capacitors are appropriate for large-scale energy storage.
2. **Ceramic capacitors** are often used in high-frequency and decoupling applications.
3. **Film Capacitors**: A broad variety of frequencies may be reliably capacitance using film capacitors.

Capacitors for ham radio applications, take into account variables like capacitance, voltage rating, and temperature stability.

Inductors

An inductor is a wire coil twisted around a core, usually magnetic, that, when current passes through, stores energy in a magnetic field. **Role in Ham Radio**:

1. **Filtering**: To stop high-frequency signals, filters employ inductors.
2. **Impedance Matching**: They are involved in the process of matching the impedance of various circuit components.
3. **Energy Storage:** When the current changes, inductors release energy that has been stored in their magnetic fields.

Types

1. **Air-Core Inductors**: A coil is coiled around a non-magnetic core in an air-core inductor.
2. **Ferrite-Core Inductors**: To boost inductance, use a ferrite core.

When choosing inductors for ham radio applications, factors including inductance, current rating, and core material are crucial.

Ohm's Law and Calculations of Power

Ohm's Law

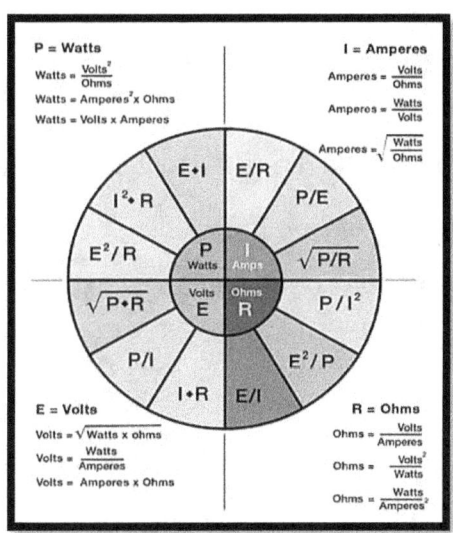

Georg Simon Ohm, a German scientist, is the name of the basic physics and electrical engineering concept known as Ohm's Law. It describes how voltage (V), current (I), and resistance (R) are related in an electrical circuit.

The mathematical expression for Ohm's Law is:

V=I·R

Where:

- V is the voltage across the circuit (in volts, V)
- I is the current flowing through the circuit (in amperes, A)
- R is the circuit's resistance, measured in ohms (Ω).

Use in Ham Radio

1. **Voltage (V):** The potential difference between antennas, transmitters, and receivers is referred to as voltage in ham radio terminology. For example, an amplifier or transceiver's power supply voltage is an important characteristic.
2. **Current:** The movement of an electrical charge across conductors is known as current (I). Understanding current is crucial for building circuits, choosing components, and guaranteeing correct performance in ham radio.
3. **Resistance** (R): Resistance is the force that opposes the current flow. In a ham radio system, electrical parts, transmission lines, and antennas all add to the circuit's total resistance.

Power Calculations

In an electrical circuit, power (∃P) is the rate at which energy is transmitted or work is completed.

The following formulae explain how power, voltage, and current relate to one another:

P=V·I

P=I2·R

P=RV2

Where:

- Power is denoted by P (watts, W).
- V stands for voltage (volts, or V).
- I is current (measured in amps, A).
- R stands for resistance.

Use in Ham Radio

1. **Transmitter Power**: To maximize communication and assure regulatory compliance, ham radio operators must have a thorough understanding of power. One essential factor is the electricity supplied to the antenna.
2. **Antenna Power Handling**: To avoid damage, it is essential to understand the power rating of antennas. Hams can choose antennas that can manage the transmitted power with the use of power calculations.
3. **Efficiency**: Amplifier and transmitter efficiency may be evaluated using power calculations. The ratio of output power to input power is called efficiency.
4. **Voltage and Current Limits**: To prevent overcharging components and circuits with voltage and current, it is crucial to understand power.

Basics of Radio Wave Propagation

An essential component of ham radio communication is radio wave propagation, and amateur radio operators may maximize their broadcasts by having a basic grasp of this process.

We'll examine the principles of radio wave propagation in ham radio here:

Electromagnetic radiation with extended wavelengths is known as radio waves. They are located between microwaves and infrared radiation in the electromagnetic spectrum. For communication, ham radio operators usually utilize frequencies between 1.8 MHz and 30 MHz.

Propagation Modes
Ground Wave Propagation

- Short journeys, usually up to a few hundred miles, are appropriate for this.

- Conductivity, frequency, and topography all have an impact on ground waves as they move over the surface of the Earth.
- Lower frequencies work better for ground wave propagation (1.8 to 3.5 MHz, for example).

Sky Wave Propagation

- Perfect for long-distance communication; also known as "**skip zone**" or "**skip**."
- High-frequency communications, such as those between 7 and 30 MHz, bounce off the ionosphere and land back on Earth.
- The efficacy of sky wave propagation is affected by the time-varying condition of the ionosphere.

Line-of-Sight (LOS) Propagation

- Suitable for comparatively short travel times.
- Without reflecting or refracting, signals move in a straight path.
- Buildings and slopes may impede the spread of LOS.

Ionospheric Layers

- The propagation of sky waves depends on the ionosphere. The D, E, and F layers are among the layers that make it up.
- Since the F layer can refract higher-frequency transmissions, it is very significant for ham radio communication. It is further separated into F1 and F2.

Maximum Usable Frequency (MUF) and Critical Frequency (fc)

Critical Frequency

- The maximum frequency at which an ionosphere layer is capable of refractive action.
- Signals traveling at frequencies above the crucial frequency may go past the ionosphere and into space.

Maximum Usable Frequency

- The maximum frequency at which ionospheric reflection enables dependable communication between two sites.

- Signals go through the ionosphere after the MUF and are not reflected to the Earth.

Sunlight Flux and Sunspots

- On the surface of the sun, spots are dark regions linked to higher solar activity.
- Sunspot counts have an impact on solar flux, which in turn influences Earth's atmospheric ionization.
- In general, increased solar flux leads to better ionospheric conditions for radio wave transmission.

Polarization

- The polarization of radio waves may be either vertical or horizontal.
- The desired communication and the properties of the propagation path influence the choice of polarization.

QRM and QSB

QSB (Fading)

- Changes in ionospheric circumstances that lead to variations in signal strength.
- Reception may be difficult when there is rapid fading.

QRM (Man-Made Interference)

- Interference from electrical gadgets or other radio transmissions.
- Appropriate antenna positioning and design may reduce QRM.

Antenna Considerations

- In the transmission of radio waves, antennas are essential.
- Signal strength and directionality are impacted by antenna design, height, and orientation.

Terrain Effects

- The propagation of radio waves may be greatly impacted by the topography around them.
- Radio waves can be reflected, refracted, or absorbed by mountains, bodies of water, and metropolitan settings.

Troubleshooting and Maintenance

Diagnosing and Fixing Common Issues

Due to the many components and possible sources of failure in radio systems, diagnosing and resolving typical problems in ham radio calls for a methodical approach.

1. **No Power or Transmission:**

Check the Power Source

- Verify that a dependable power supply is connected to the radio. Inspect for damage to power cables, connections, and fuses.
- Check to make sure the power supply is operating properly.

Battery Condition

- If you're using a battery, make sure it's charged enough and replace or recharge it as needed.

Transmitter Settings

- Verify that the frequency and mode on the transmitter are right.
- Inspect for any power-saving mechanisms that may be preventing transmission.
2. **Low or Absence of Signal Reception**

Antenna Connection

- Check for loose or broken connections on the antenna connection.
- Examine the antenna for changes in location or damage.

Antenna Tuner

- Make sure the antenna tuner is correctly adjusted to match the frequency if you're using one.

Receiver Settings

- Verify that the frequency, mode, and bandwidth are set correctly on the receiver.
- Check that any attenuators or filters are adjusted correctly.

Interference

- Locate and remove any possible sources of interference, such as power wires or adjacent electrical devices.

3. **Audio Issues:**

Speaker and Microphone Connections

- Check speaker and microphone connections for frayed or broken wires.
- Make sure the microphone on another radio or gadget is working properly.

Gain and Volume Configurations

- Set the transmitter and receiver's volume and gain settings to the proper values.
- Look for any controls that are off or muted.

4. **Transmission/Reception Intermittent**

Loose Connections

- Verify that all connections—cables, connectors, and grounding points—are tight.
- Take extra care with any connections that could have come loose as a result of vibration.

Faulty Components

- Examine the circuit board for any malfunctioning resistors, capacitors, or transistors.
- Examine visually for indications of corrosion, burn marks, or damaged parts.

5. **Frequency Drift**

Stability Settings

- Modify the radio's frequency stability settings, particularly if you're using outdated gear.
- Before using the radio, make sure it has had enough time to warm up.

Antenna and Feedline

- Check to make sure the feedline and antenna are correctly matched and not causing frequency drift.

6. **RF Interference/Feedback:**

Isolation

- To shield wires from RF interference, add ferrite beads.
- Make sure all equipment is properly grounded to minimize the possibility of feedback.

Location of Antenna

- Verify where the antenna is positioned to reduce the possibility of feedback from surrounding buildings.

Filtering

- Put band-pass filters in place to reduce interference from nearby frequencies.
7. **Problems with Digital Modes:**

Sound Card Configuration

- Check the digital modes on the sound card.
- Verify that the radio and computer have the proper interface for digital communication.

Software Configuration

- Verify that the software settings for digital modes correspond to the radio's requirements.
- Install the most recent firmware and software updates.
8. **Standing Wave Ratio Issues (SWR):**
- If you're using an antenna tuner, make sure it's adjusted for a low SWR.
- If SWR remains high, look for a malfunctioning feedline or antenna.
- Examine the antenna visually for any indications of wear and tear or structural alterations.
- To reduce SWR, make sure the antenna equipment and radio are properly grounded.

Periodic Inspections and Preventive Maintenance

Preventive maintenance and routine inspections are essential to keeping a ham radio station dependable and operational. Proactive maintenance and routine inspections assist in guaranteeing that equipment runs smoothly, lower the likelihood of malfunctions, and increase the equipment's lifetime.

Regular Inspections

Antennas

- Examine antennas for corrosion, physical damage, or loose connections.
- Examine connections, coaxial cables, and insulators for wear indicators.
- Check to make sure the antenna is secured firmly and is appropriately grounded.

Transceivers

- Inspect transceivers for physical damage, loose knobs, or damaged screens.
- Inspect and sanitize switches and connections.
- Look for charred or discolored interior parts as indicators that something is overheated.

Power Supplies

- Check the connections and power cords.
- Listen for odd sounds, heat exhaustion, or burned odors.
- Check that the output voltage is within the prescribed range by measuring it.

Coaxial Cables

- Check for damage, kinks, or wear on coaxial wires.
- Inspect connections for corrosion and tightness.
- Check that the cable is continuous to make sure there are no breaks in it.

Grounding Systems

- Check for corrosion in grounding systems and make sure the bonding is correct.
- Verify the integrity and tightness of ground connections.
- Confirm that the grounding system complies with the local electrical code.

Masts and Towers

- Check the structural integrity of the towers and masts.
- Check for physical damage such as loose bolts, corrosion, or other issues.
- To guarantee smooth functioning, lubricate moving components, such as rotators.

Preventive Maintenance

Cleaning

- Regularly clean the equipment using the proper instruments and supplies.
- Clean transceivers, antennas, and other parts of any dust, dirt, or other impurities.
- To clean spots that are difficult to reach, use a vacuum or compressed air.

Lubrication

- To reduce wear and friction, lubricate moving elements including rotators and tuning mechanisms.
- To guarantee compatibility and efficacy, use lubricants as suggested by the manufacturer.

Temperature Control

- To avoid equipment overheating, make sure it has enough airflow.
- When necessary, clean or replace the air filters in the cooling fans.
- Keep equipment away from heat sources and direct sunshine.

Software Updates

- Frequently check transceivers and other digital devices for firmware and software upgrades.
- To improve performance and fix any vulnerabilities, update software under the manufacturer's instructions.

Backup Systems

- Make regular backups of setups, code, and crucial data.
- Keep backups safe to enable speedy recovery in case of equipment loss or malfunction.

Software and Firmware Upgrades

For ham radio equipment to continue operating at peak efficiency, to add new features, and to remain compatible with emerging technology, firmware and software upgrades are essential. The procedure includes upgrading the external software, which might be installed on a computer or mobile device, as well as the internal firmware, or programming, of the radio's microcontroller.

Here's what you should do:

- Find out what brand and model your ham radio is. Firmware and software updates may be handled differently by different manufacturers.
- Check for firmware and software upgrades by going to the manufacturer's website. New versions are often released by manufacturers to fix issues, improve performance, or add new features.
- Visit the official website of the manufacturer to get the most recent firmware and software files. Make sure the files you are downloading are the right ones for the particular radio model you possess.
- It is essential that you back-up the existing settings on your radio before starting with the update. Usually, you may do this by utilizing the radio's menu system or specialized programming tools.
- Go over the release notes that the manufacturer sent. The updates and enhancements made to the latest firmware or software version will be detailed in this document. Having a clear understanding of these modifications will enable you to decide whether or not to go forward with the update.
- Turn off the radio's power and use the proper interface connections to connect it to a computer. Certain radios may link via serial, USB, or other interfaces. For correct connection, adhere to the manufacturer's recommendations.
- If you want to upgrade the firmware on your particular radio model, follow the manufacturer's instructions. Using a firmware update tool on your computer while

the radio is attached may be necessary to do this. Try not to disconnect the radio during the update process, and be patient.

- To install the updated software version on a computer, launch the setup file you downloaded. To complete the installation procedure, adhere to the on-screen directions. Software may be used by some radios to program memory, change settings, or carry out other tasks.
- Power cycle the radio when the update is finished to make sure everything is working as it should. Verify the accuracy of the settings and configurations by checking them.
- Restore your stored radio settings if needed. To save your customized setups, you must take this action.
- Test the radio to make sure it's functioning properly after the update. Look for updates, enhancements, or any problems that could have been fixed.
- Think about contacting the manufacturer if you have any problems or comments about the update procedure. They may use this to enhance their support and product offerings going forward.

Frequently Asked Questions

1. What are the different Ham Radio licensing classes?
2. How do you prepare yourself for the Ham Radio exams?
3. What is the Ohm's law and calculations of power in Ham Radio?
4. How do you upgrade your Ham Radio license?

CHAPTER FOUR
ADVANCED OPERATING TECHNIQUES

Overview

Want to go deeper into Ham Radio? Chapter Four talks about the various advanced operating techniques in Ham Radio including Contesting and DXing, satellite operations, and others.

Contesting and DXing

Participating in Amateur Radio Contests

For radio aficionados, taking part in amateur radio competitions may be a fulfilling and thrilling experience. The use of allocated radio frequencies for non-commercial communication, experimentation, and public service is known as amateur radio or ham radio. In the amateur radio community, contests provide operators a chance to demonstrate their abilities, test their gear, and make connections with other enthusiasts throughout the globe.

Understanding Amateur Radio Contests

Types of Contests

- **DX Contests:** Concentrate on establishing contact with operators who are located far away.
- **Field Day**: highlights functioning in field circumstances and emergency preparation.
- **RTT Contests:** Communicate by radio teletype.
- **CW Contests**: These mostly include communicating via Morse code.
- **Phone Contests:** Use voice modes (AM, FM, SSB, etc.) in phone contests.

Contest Calendar

- Examine the contest schedules offered by groups such as the Amateur Radio Relay League (ARRL) and others.
- Take note of each contest's date, time, and regulations.

Preparing for a Contest

Equipment Check

- Verify that the radio equipment you own is in proper operating order.
- Check that antennas are tuned for the contest frequency bands by testing them.

Logging Software

- Select logging software and get acquainted with it to ensure precise record-keeping.
- Make sure your program supports the logging formats required by the several competitions.

Become acquainted with the rules

- To prevent being eliminated from the contest, read and understand the regulations.
- Be mindful of the permitted operating modes, interchange formats, and scoring systems.

Operating during the contest

Band and Mode Strategy

- Select the bands and settings according to your equipment and level of experience.
- Keep an eye on the circumstances of propagation and change bands as necessary.

Effective Logging

- To prevent misunderstanding, quickly log every interaction.
- Provide all necessary data, such as exchange details, serial numbers, and signal reports.

Pacing Yourself

- Given the length of certain contests, pace yourself to prevent weariness.
- Take pauses to recover and stay focused.

Contest Manner

- Comply with contest protocol, which includes refraining from making repeated calls on a certain frequency.

Chasing DX (Distant) Stations

One of the most exciting aspects of amateur radio (ham radio) is chasing DX (distant) stations, which is contacting radio stations that are in uncommon or far-off areas. DXing, as it is generally referred to, is sought by ham radio operators all over the globe and adds an intriguing depth to the game.

This is a comprehensive guide on ham radio DX station chasing:

Fundamentals of DXing

1. **DXCC**: The American Radio Relay League (ARRL) oversees the DXCC program, which honors amateur radio operators who have effectively communicated with a certain number of organizations or countries. From the basic DXCC award to more significant honors like the Honor Roll or 5-Band DXCC, DXers strive to acquire a variety of levels.
2. **Entities and Countries**: In ham radio, an **"entity"** is a political or geographic place. Since each entity is tallied independently, DXers often aim to get in touch with as many entities as they can.

3. **Propagation and Bands:** The present ionospheric conditions, which impact radio wave propagation, affect the ability to establish contact with DX stations. Operators may move between bands based on the time of day and meteorological conditions since various bands (frequency ranges) display distinct propagation characteristics.

Tools and Antennas

1. **High-frequency (HF) Bands**: 20 meters, 15 meters, and 10 meters are examples of HF bands where DXing is most common. Transmission over large distances is possible in these bands.
2. **Directional Antennas:** Directional antennas, like Yagi antennas, are used by many DXers to concentrate their signals in a particular direction. This makes it easier to target distant stations.
3. **Amplifiers**: When attempting to contact stations in difficult-to-reach places or tough weather, DXers often employ amplifiers to boost the strength of their signals.

Methods for Chasing DX

1. **Listening abilities**: To distinguish faint signals from background noise, skilled DXers use their acute listening abilities. To handle pile-ups, they often use split frequency operation, in which stations listen on one frequency and broadcast on another.
2. **Pile-ups:** When a rare DX station is broadcasting, many operators try to connect with it at the same time, which results in a pile-up. Competent operators can maneuver past pile-ups with efficiency.
3. **Propagation Predictions**: Solar activity, ionospheric conditions, and other elements influencing radio wave propagation are monitored by DXers. Real-time data and forecasts are available via online tools and services to help in DX contact planning.
4. **QSL Cards**: QSL cards are often used to confirm a successful DX contact. These cards are used as a means of communication confirmation between operators. As a memento of their accomplishments, many DXers collect these cards.

Practicing Ethics

1. **Respect Operating rules:** To guarantee impartial and courteous communication, DXing has set up operating rules. These include habits like not making a lot of calls, listening before sending and paying attention to the DX station's directions.
2. **QRM (Interference):** It's important to minimize interference and refrain from interfering with other stations. Operators preserve a good working environment by following the rules of frequency etiquette.

Challenges

1. **Weather and Geographical Difficulties**: When attempting to contact certain DX entities, weather conditions, geographic distance, and local laws may provide difficulties.
2. **Competitive Environment**: There might be fierce rivalry among operators as they compete for the same uncommon DX station. The secret to success is to be persistent, and patient, and to follow best practices.

Strategies for Successful Contesting

In amateur radio contests, operators, or "**hams**," attempt to make as many connections as they can in a certain amount of time. This is known as contesting. Strategic preparation, efficient station setup, and technical proficiency are all necessary for successful competition.

The strategies include the following:

- Recognize how the particular contest you are entering operates in terms of rules and scoring. There are differences in the exchange forms, time limits, and scoring systems across competitions.
- Check that all of your equipment—transceivers, antennas, and power supply, among others—is operating at peak efficiency. Make sure everything is in working order by doing a comprehensive inspection.
- Become familiar with the contest's band plan and propagation circumstances. Create a frequency strategy that will increase the likelihood that you will establish contact on several bands and modes.

Station Setup

- Make sure your antennas are appropriate for the bands and modes you want to use. To maximize signal strength and reduce interference, rotate directional antennas.
- Use resources like cluster networks, logging software, and propagation forecast tools to keep tabs on band conditions, follow other stations, and manage your connections effectively.
- Keep backup gear and power supplies on hand in case of malfunctions. During a competition, redundancy is essential for uninterrupted functioning.

Operational Techniques

- To make the logging process go more quickly, use computer logging software. As per the contest rules, make sure your log has all the information needed for each interaction.
- Choose between being a "**run**" operator, holding a frequency, and calling CQ to draw in more stations, or a "**search and pounce**" operator, looking for and contacting stations. Adapt your plan according to the available contacts and the band circumstances.
- Recognize peak operation hours, propagation circumstances, and band openings. Maintain a healthy pace to prevent weariness and carefully schedule breaks to enhance productivity.

Communication Skills

- Make sure all of your communications are succinct and clear. To optimize contacts, use standardized phonetics and provide information quickly.
- When speaking on the radio, have a courteous and professional manner. Observe the rules about contested behavior and show consideration for other operators.

Adaptability

- Keep an eye on the band conditions and be ready to move to another one if the propagation circumstances change. Make use of many bands to boost the rate of interaction.

- Master many modes, particularly if the competition permits it. The ability to transition between digital, SSB, and CW modes may increase your contact prospects.

Post-Contest Analysis

- Make sure you submit your log under the contest rules. To improve your score, turn in your log as soon as possible and precisely.
- Evaluate your performance after the competition. Examine what went well and pinpoint areas that need improvement. To improve your contending abilities for the next tournaments, take lessons from your past experiences.

Satellite and Moonbounce Operations

Setting Up for Satellite Contacts

In ham radio, setting up satellite contacts may be a fun and fulfilling experience. Satellites allow amateur radio operators, or **"hams,"** to communicate with people all over the world, extending the range of their transmission beyond conventional VHF and UHF bands.

Tools Required

- **VHF/UHF Transceiver:**

Select a transceiver that can operate in full duplex on both bands (VHF and UHF). Transmitting and receiving concurrently is made possible using full-duplex technology, which is essential for satellite communication.

- **Antennas**:

The most typical configuration is a pair of crossed yagi antennas, one for UHF and one for VHF. The satellite should be tracked by mounting these antennas atop a rotator.

- **Rotator System:**

Invest in an elevation and azimuth rotator system so you can follow the satellite as it travels through space. While some hams still use manually adjusted antenna systems, tracking is made much easier with automatic rotators.

- **Satellite Tracking Software**:

Use satellite tracking software to ascertain the angles and pass times specific to your position. SatPC32, GPredict, and Heavens-Above are a few examples. You may use these tools to find out when and where you can see a satellite from where you are.

- **Rig Control Interface**

Use a computer with rig control software to connect your transceiver. This gives the computer the ability to adjust your radio's frequency and mode when a satellite is passing above.

- **Headphones**

If you want to improve your ability to detect weak signals, particularly during low-altitude passes, use noise-canceling headphones.

- **Portable Power Source**

Make sure you have a dependable portable power source, such as batteries or a portable power generator, if you want to work in the field.

- **Satellite Pass Predictions**

Familiarize yourself with websites or applications that offer real-time satellite pass forecasts for your location before you begin. This will help you arrange your operational hours.

Set up Procedures

1. Mount your VHF and UHF antennas on the rotator system and make sure they are polarized correctly. This is the first step in the antenna setup procedure. Yagi antennas that are crossed are often used to capture both horizontal and vertical polarization.

2. Enter the specifics of your position into the program developed for tracking satellites. You will need to connect it to your transceiver and then configure the tracking settings that are required.

3. Use a rig control interface to establish a connection between your transceiver and the computer. You will need to configure the software so that it can alter the frequency and mode of your radio depending on the specifics of the satellite pass.

4. It is recommended that you carry out a test before the pass to guarantee that your rotator system, tracking software, and rig control interface are all functioning faultlessly together.

5. Listen to the downlink frequency of the satellite, and then use your rotator system to track the satellite after it has been located. When you want to contact other hams via the satellite, you need to transmit on the uplink frequency.

6. Maintain a diary of your satellite contacts, making sure to record the call signs of the stations you interact with as well as the satellite name, frequency, mode, and time each.

7. You should regularly upgrade the software and firmware of your satellite tracking system, and you should also practice operating the system to improve your abilities and optimize your setup.

Utilizing the Moon for Long-Distance Communication

A technique known as Earth-Moon-Earth (EME) communication is an interesting concept that includes utilizing the Moon as a natural satellite to bounce radio signals. This method is also known as EME communication. By using this technology, hams can communicate over extraordinarily great distances, beyond the restrictions of conventional terrestrial communication.

Key Concepts

Earth-Moon-Earth (EME) Communication

- Electromagnetic (EM) communication, often known as **"moonbounce,"** is a method of communication that involves transmitting radio signals from Earth to the Moon and then receiving the reflected signals that are sent back to Earth.
- Although it does not possess any power source of its own, the Moon functions as a passive reflector, reflecting communications. To overcome the difficulties associated with poor signal strength, this method calls for the use of high-gain antennas and advanced equipment.

Antenna Considerations

- High-gain antennas are necessary for EME communication owing to the lengthy round-trip journey of signals to and from the Moon.
- Yagi antennas with many elements, big dish antennas, or customized EME arrays are often employed to provide the needed gain.

Frequency Selection

- EME communication commonly happens in the microwave and higher frequency bands, notably in the VHF, UHF, and SHF ranges.
- Frequencies around 144 MHz, 432 MHz, and even higher bands are common candidates for moonbounce transmission.

Transmitter Power and Receiver Sensitivity

- Due to the feeble signals reflected by the Moon, great transmitter power and sensitive receivers are needed.
- Amplifiers capable of delivering several hundred watts or more are commonly used combined with low-noise preamplifiers to amplify weak signals on the receiving end.

Propagation Delays and Doppler Shift

- Signals going to and from the Moon have propagation delays, which must be corrected for in equipment design.
- Doppler shift, induced by the relative motion of the Earth and the Moon, needs modifications to ensure frequency accuracy during transmission.

Operating Procedures

- EME communication needs precise coordination between transmitting and receiving stations.
- Operators typically depend on specified timetables, utilizing exact timing to improve communication during times of optimum moon visibility.

Software Aids

- Various software tools aid operators in forecasting good EME circumstances, including Moon monitoring software and prediction tools.
- These tools assist in determining the ideal periods for communication depending on the Moon's location, avoiding wasted efforts during adverse circumstances.

Tracking Satellite Passes and Orbits

Tracking satellites and their passes is an intriguing feature of amateur radio (ham radio). Numerous satellites routinely orbit the Earth, including those used for amateur radio, communication, and weather monitoring. One further level of difficulty and enjoyment to the ham radio activity is tracking these satellites and establishing contact with them.

Understanding Satellite Orbits

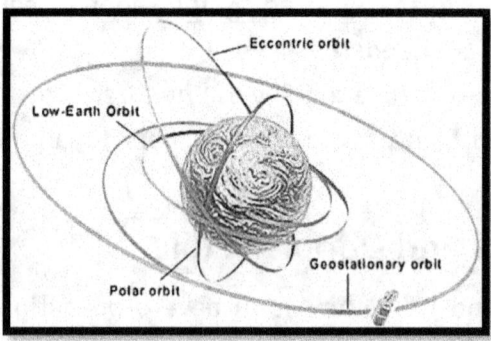

Different Orbit Types

- **Low Earth Orbit (LEO):** Satellites in Low Earth Orbit (LEO) travel between 180 to 2,000 kilometers above the planet. Most amateur radio satellites are in low Earth orbit (LEO).
- **Medium Earth Orbit (MEO):** The Medium Earth Orbit (MEO): GPS satellites and other navigation satellites often use this orbit.
- **Geostationary Orbit (GEO):** Typically employed for communication satellites, GEO orbits place spacecraft at a fixed location concerning the surface of the Earth.

Keplerian Elements

- Keplerian elements, which include variables like inclination, eccentricity, and semi-major axis, are used to characterize the orbits of satellites. These components are used by software programs to precisely estimate the satellite's location.

Tracking Satellite Passes

Predictive Tools

- **Heavens-Above**: A well-known web application that, depending on your location, forecasts when observable satellite passes will occur.
- **N2YO.com:** Provides estimates for impending passes along with real-time monitoring of satellites.

Tracking Software

- **Gpredict:** Real-time monitoring of satellite passes and scheduling of upcoming observations are made possible by this open-source satellite tracking program.
- **Orbitron:** An additional free satellite tracking program with an easy-to-use UI.

Ham Radio Equipment and Transceivers

- To transceive and receive concurrently, use a transceiver that supports full-duplex operation.
- Take into consideration the Doppler shift, which modifies the signal's frequency when the satellite moves in or out of the observer's line of sight. Doppler shift parameters are modifiable on a lot of contemporary transceivers.

Special Event Stations

Organizing and Participating in Special Events

In the world of ham radio, planning and taking part in special events may be thrilling and fulfilling. Amateur radio operators may get together, show off their talents, and commemorate special moments at special events. Effective preparation and organization are essential for every event, whether it is a competition, memorial service, or volunteer project, to ensure its success.

Organizing Special Events

- Clearly state the special event's objective. It may be a competition, an anniversary celebration, a demonstration in public, or a practice for emergency communication.
- Decide on a time and date that will best serve the goal and enable the greatest number of participants. Verify whether it coincides with any other significant ham radio events.
- Choose whether the event will operate from a mobile location or at a permanent site. Think about things like infrastructure availability, safety, and accessibility.
- Acquire any licenses or authorization required for the selected site. In the case that public areas are involved, work with the local government to make sure that all rules are followed.

- Choose the event's structure, including whether it will be a social event, a competitive competition, or a display of emergency communication skills.
- Spread the word about the event among ham radio operators via radio clubs, social media, and newsletters. If the event is public, get in touch with the local media to reach a larger audience.
- Look for sponsorships from companies that could be interested in sponsoring the event, such as radio equipment makers or nearby businesses.
- Consider practical matters like parking, electricity, shelter, and emergency medical assistance.
- Find and organize volunteers to fill a variety of positions, such as public relations reps, operators, loggers, and event planners.
- Verify that every piece of essential radio equipment is operational. Plan for backup equipment in case anything goes wrong.
- Create the necessary paperwork, such as operating procedures, frequency allocations, and event rules. Provide participants with this information ahead of time.

Taking Part in Special Events

1. Make sure you are familiar with the operational procedures, rules, and specifics of the event. Before use, test your equipment to find any problems.
2. To prevent interference, stick to assigned frequency ranges. Observe appropriate procedures and pay attention to band plans.
3. Keep precise records of the people you contact. Once the incident has occurred, instantly submit your record by following the prescribed logging protocols.
4. Follow established procedures if the event incorporates emergency communication drills. Develop your ability to communicate clearly and succinctly.
5. Be ready to engage with the public if the event is open to the public. Answer inquiries, provide information on amateur radio, and highlight the features of ham radio.
6. Keep proper operating protocol in place. Show sportsmanship, be polite on the air, and stay away from needless interruptions, particularly during competitive activities.

7. Share your experiences with the ham radio community after the event. Comment on what went well and provide suggestions for how to make future events even better.

QSL Cards and Confirming Contacts

QSL cards are essential for verifying and recognizing connections made in the amateur radio (or "**ham**") community. These cards are a classic way to verify the specifics of a contact and provide concrete evidence of communication between two amateur radio operators. The Q code is a system of three-letter codes used in radio communication to send brief messages. The name "**QSL**" is derived from this code.

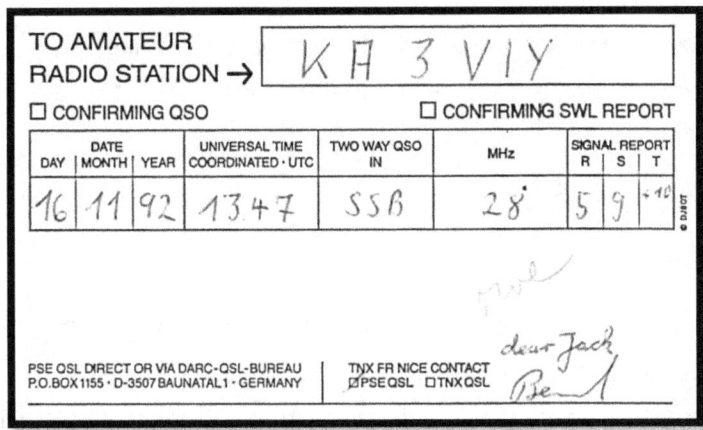

Here's a thorough explanation of QSL cards and how ham radio interactions are verified:

QSL Cards

Purpose

- To verify the specifics of a two-way conversation (QSO) between amateur radio operators, QSL cards are used.
- They include information on the stations engaged, the day and time, the frequency, and the manner of communication. They also serve as confirmation that a contact has occurred.

Design

- QSL cards are available in a variety of styles and layouts, from simple to complex.

- The call signs of both stations, the contact date and time, the frequency used, the communication method (voice, Morse code, digital), and a section for signal reports are examples of common aspects.

Exchanging QSL Cards

- To verify the specifics of the conversation, operators may trade QSL cards after a successful connection.
- Not all operators engage in this voluntary exchange. Nonetheless, it is a highly valued and often used feature among the amateur radio community.

Logbooks and electronic QSL (eQSL)

- Apart from physical cards, electronic QSLs have gained popularity. Operators may exchange confirmations electronically using websites like eQSL.cc.
- Contacts are often tracked using logging software, and some programs can automatically create and send eQSLs.

Confirming Contacts

Logging

- To document every contact's specifics, such as call signs, date, time, frequency, and signal reports, amateur radio operators keep logbooks.
- Verifying connections and taking part in competitions and rewards need logging.

Checking and QSL Bureau

- Operators have the option of mailing or using a QSL bureau to confirm a contact.
- An organization that makes it easier for amateur radio operators to exchange QSL cards is known as a QSL bureau. Postage expenses are minimized by regularly gathering and distributing cards.

ARRL's Logbook of the World (LoTW)

- The online service Logbook of the World is offered by the American Radio Relay League (ARRL).
- If both parties in a contact use LoTW, operators may electronically verify confirmations by uploading their records.

Frequently Asked Questions

1. How do you set up satellite contacts?
2. How do you participate in amateur radio contests?
3. How do you organize and participate in special events?
4. What are the strategies for successful contesting?
5. How do you use the moon for long-distance communication?

CHAPTER FIVE

DIGITAL MODES AND SOFTWARE

Overview

In this chapter, you will learn everything about digital Modes and software in Ham Radio including digital communication, packet radio, APRS and so much more.

Introduction to Digital Communication

FT8, PSK31, and Other Digital Modes

In ham radio, the term "**digital modes**" refers to techniques for sending and receiving data that make use of digital signals as opposed to conventional analog communications. Effective and dependable communication is made possible by these modes, particularly in difficult situations. Here, we will examine two widely used digital modes in ham radio, FT8, and PSK31, as well as briefly discuss a few more prevalent digital modes.

FT8 (Franke-Taylor design, 8-FSK modulation)

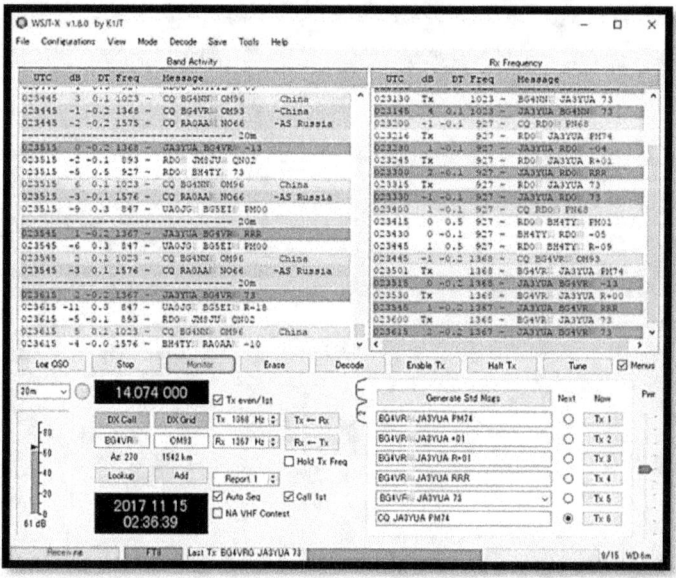

FT8 uses an 8-FSK (Frequency Shift Keying) digital modulation scheme, in which various frequencies stand for various binary values. Since FT8 transmissions are short—only 15 seconds—they are ideal for low-power and weak signal communications. FT8 uses robust forward error correction, which enables it to decode signals under difficult propagation circumstances.

How it works:

1. **Message Format**: Grid squares, callsigns, and signal reports are all included in structured FT8 messages.
2. **Automation:** Software manages the majority of the transmission and reception operations in FT8, which is a highly automated system.
3. **QSOs (Conversations):** FT8 is perfect for initiating connections in busy band settings since it allows for quick information exchange.

WSJT-X and JTDX are two commonly used programs for FT8.

Phase Shift Keying 31 Baud, or PSK31

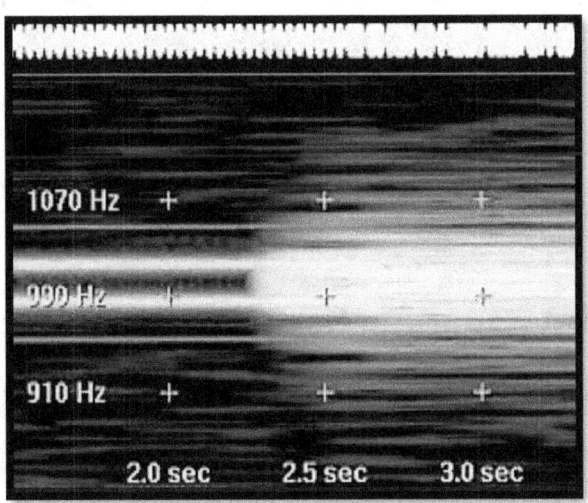

PSK31 modulation uses phase shift keying, in which shifts in phase correspond to distinct symbols. This protocol is appropriate for busy bands and very narrow bandwidths because of its very small bandwidth (about 31 Hz) and symbol rate of 31.25 baud. Since PSK31's error correction is weak, it could not function effectively in noisy environments.

How it works:

1. **Typing Interface:** PSK31 functions with a basic typing interface where the operator enters messages on the keyboard.
2. **Real-Time Communication:** It enables real-time communications, much like a chat, with the operator viewing the received text on their screen as it is received.

Popular software for PSK31 includes Fldigi and Digipan.

Other Digital Modes

RTTY (Radio Teletype)

RTTY uses frequency shift keying to represent characters. It's a heritage mode that has been in use for a long time, and some hams still love using it.

JT9 and JT65

Comparable to FT8: These modes, which have slower transmission rates than FT8, are also intended for poor signal communication and were created by the same authors.

Packet Radio

Using ham radio frequencies, packet radio allows digital communication between computers. It facilitates file and message sharing and supports BBS (Bulletin Board System) functions.

Automated Packet Reporting System, or APRS

Real-time location and other data communication is facilitated by the use of APRS. Frequently utilized for weather stations, emergency communication, and car tracking.

Advantages and Challenges of Digital Modes

Benefits of Digital Ham Radio Modes

1. Compared to conventional analog modes, digital modes use the available bandwidth more effectively. Within the same frequency range, more simultaneous communications are possible thanks to this efficiency.

2. In difficult circumstances like poor signal propagation or loud background noise, digital modes frequently offer higher signal quality. As a result, communication is more trustworthy and clear.

3. Several digital modes use error correction techniques to guarantee that transmitted data can be accurately received even in the presence of noise or interference. This is especially helpful when there is a weak signal.

4. A substantial amount of information can be transmitted in a compressed format using digital modes. This makes data transmission faster and more effective for a range of uses, including emergency communications.

5. Text and pictures may be sent over the radio thanks to digital forms of transmission capability. Due to their adaptability, they are useful for providing weather maps, SSTV (Slow Scan Television), and other types of data in addition to voice communication.

6. This allows for the use of software applications and interfaces, which is made possible by the smooth integration of digital modes with computer technology. By providing ham radio operators with features such as automated logging, frequency management, and digital signal processing, this integration provides an enhancement to the capabilities of ham radio operators.

7. The ability to communicate in an energy-efficient manner is made possible by certain digital modes that are engineered to function well at lower power levels. In situations when power saving is of the utmost importance, such as portable and emergency settings, this is useful.

8. Digital modes often adhere to established protocols and modulation schemes, which enables compatibility between various radio equipment and makes it easier to communicate across international borders. Standardization on a worldwide scale helps to ensure uniformity and makes things easier to utilize.

Challenges

1. The learning curve for digital modes may be steeper than the learning curve for standard speech modes when entering the digital mode. Operators must possess a comprehensive understanding of certain software, hardware interfaces, and digital protocols.

2. The implementation of digital modes could need extra hardware, such as interfaces for sound cards, digital signal processors, and programs that are specifically designed for the purpose. The total cost of entering the digital ham radio market may rise as a result of this.

3. It is possible that digital modes, although being more economical with bandwidth, nonetheless have restrictions in terms of the amount of bandwidth that is accessible. Congestion on specific bands may become a problem as more people use digital alternatives.

4. In digital modes, computers and the gear that goes along with them are very important. Communication may be disrupted due to technical faults with these components or software errors, but this may not be the case with older analog means.

5. The performance of digital modes may be negatively impacted by the presence of other electronic equipment as well as interference from radio frequency emissions. It is also possible that not all ham radio operators utilize equipment that is compatible with one another or adheres to the same standards, which might result in compatibility difficulties.

6. Some countries or regulatory organizations have certain limits or requirements for the use of particular digital modalities. Ham radio operators are required to be aware of these restrictions and to maintain compliance with them.

7. While digital modes are vulnerable to some types of interference and hacking, analog modes are not. The operators are obligated to handle the issue of ensuring the security of the data that is being transferred, particularly in sensitive applications.

8. Under some propagation circumstances, such as severe fading or multi-path effects, digital modes may not always function at their ideal level. For communication to be trustworthy, it is vital to have a solid understanding of how digital signals behave under a variety of conditions.

Software Tools for Digital Operation

Digital Mode Software

- **Fldigi**: Fldigi is a well-known piece of open-source software that is capable of supporting a wide range of digital modes, including PSK, RTTY, Olivia, and many more. To encode and decode digital signals, it offers a straightforward user interface.

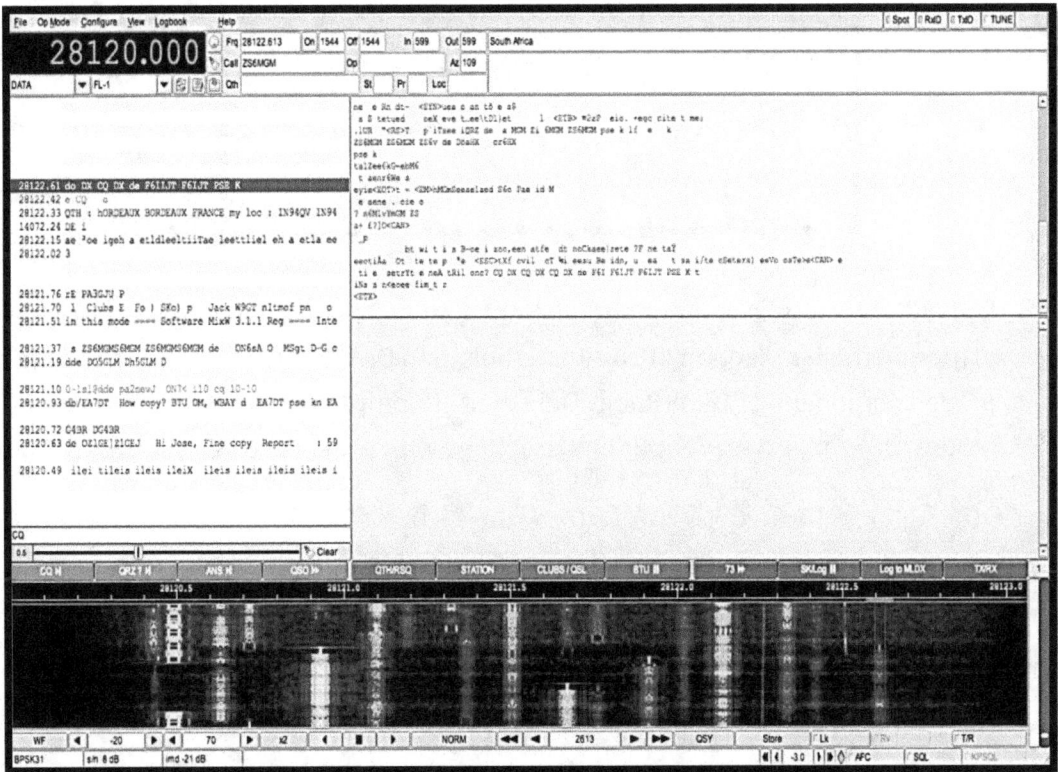

- **WSJT-X**: This program, known as WSJT-X, was developed with the express purpose of facilitating communication over weak signals, notably for modes like FT8, FT4, and JT65. To facilitate communication under difficult circumstances, it makes use of advanced digital signal processing methods.

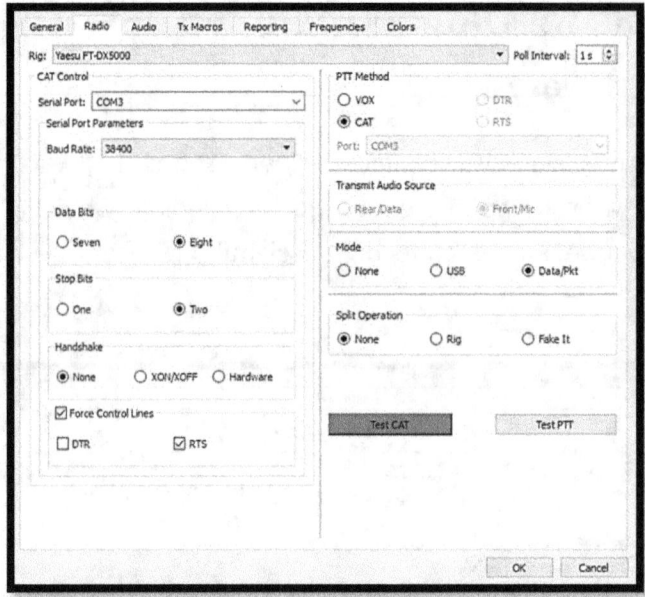

- **DM780**, which is a component of Ham Radio Deluxe: Ham Radio Deluxe is a comprehensive collection of software tools, and DM780 corresponds to the digital mode component of Ham Radio Deluxe. It is compatible with a broad variety of modes and integrates without any problems with other features such as logbooks and rig control software.

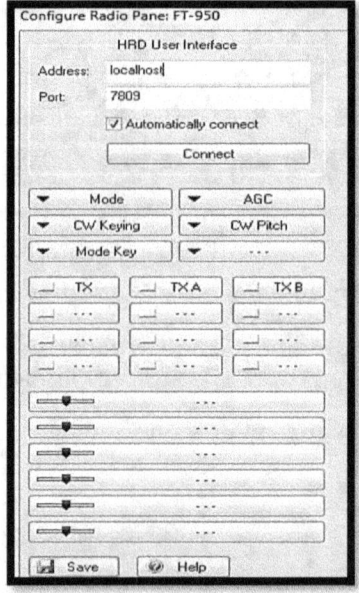

Logging Software

- **Logger32**: Logger32 is all-encompassing logging software that is compatible with a wide range of radios and digital mode applications. It offers capabilities for handling awards, keeping track of contacts, and creating reports using the information.
- **N1MM Logger+:** This program was developed primarily for contest logging. It is compatible with digital modes and offers features like real-time scoring, multi-operator setups, and the ability to interface with digital mode software.
- **CQRLOG** is a logging tool that contains capabilities for maintaining awards, handling QSOs, and connecting with digital mode programs. CQRLOG is also known as CQRLOG. Several different types of rig control interfaces are supported by it.

Rig Control and CAT Software

- **Ham Radio Deluxe (HRD):** In addition to its capabilities for logging and digital mode, HRD also includes facilities for rig control. It is compatible with a wide variety of well-known transceivers and gives users the ability to operate their radios directly from their computers.
- **Hamlib** is a library that offers a standardized interface for rig management. Hamlib is a common name for this library. Hamlib is used by a multitude of logging and digital mode applications to communicate with a variety of transceivers.

Digital Signal Processing (DSP) Software

- **FLDigi (built-in DSP):** FLDigi is equipped with built-in DSP capabilities, which allow it to filter and process digital signals. People can customize their reception to suit certain modes and settings thanks to this.
- **GNU Radio** is an open-source toolkit for the construction of software-defined radios. It is designed for people who are interested in more complex signal processing and experimentation. This software offers a graphical user interface to build flowgraphs for signal processing.

Antenna Modeling Software

- **EZNEC** is a tool that is often used for antenna modeling, even though it is not precisely a digital mode tool. Hams can improve their antenna systems' performance over a wide range of bands and situations with the assistance of this tool.

APRS and Packet Radio

A digital communication system known as the Amateur Packet Reporting System (APRS) is deployed by amateur radio operators to allow for the exchange of messages and the tracking of positions via the use of GPS. With multiple uses in tracking, emergency communications, and weather reporting, APRS is a well-known technology among hams.

What is the Amateur Packet Reporting System (APRS)?

To exchange messages and information over short distances, the Advanced Packet Radio Service (APRS) is a digital communication system that transmits data packets using radio waves. During the 1990s, Bob Bruninga, who is also known as WB4APR (SK), was the one who first invented it. Since then, it has gained widespread use among amateur radio operators. APRS can function by sending digital packets of information, such as GPS coordinates, to other stations that are within its range. After that, the packets are sent to the APRS network, which is made up of several gateways and digipeaters that are connected over the Internet. Additionally, the International Space Station is equipped with an APRS digipeater, which is capable of relaying packets when the station is in the position above. The Automatic Positioning and Response System (APRS) have a broad variety of applications, including the monitoring of the locations of weather stations, vehicles, and people.

Setting up APRS

Numerous pieces of hardware and software are used in the process of APRS configuration.

Examples of them might be:

- A radio that can use APRS built right in. Handheld devices and mobile rigs that are capable of generating packets and receiving them are included in this category. In the North American region, the radio is commonly tuned to run at 144.390 MHz.
- APRS signals may also be generated by a computer using specialized software, and then they can be sent to a radio that does not have APRS capability. The program that is used in this circumstance is either "**MacAPRS**" or "Dire Wolf (Decoded Information from Radio Emissions for Windows Or Linux Fans)." After that, a sound card interface is utilized to connect the radio to the computer.
- The mobile phone app for Apple and Android devices is available. Although these do not make use of the radio frequency (RF) components of APRS, they can include position information into APRS system packets by using their GPS and data connections.

Here are the steps:

1. **Configure the radio**
- Radio with the capacity to use APRS: Configure the software under the instructions provided by your radio.
- Installation of APRS software on your computer and the creation of an audio interface between your personal computer and radio are both components of radio and computer design. For straightforward reception, a cable that connects the audio output of the radio to the line input of the personal computer will

suffice. Install a version of the program that may meet your requirements or one of the versions that were specified above.

2. In North America, the terrestrial network should have the packet frequency set to 144.390 MHz on the frequency setting.

3. **Software Configuration**

- You need to make sure that the route is typed into the radio that has APRS functionality. Receiving stations are informed of the number of times they should digipeater your packet via this. A common example of a setting is WIDE2-2.

- Configure the APRS software so that it is compatible with your radio by adjusting the frequency, baud rate, and audio levels. This is part of the radio and computer design process. You can receive a password for your callsign on the internet if you want your computer to communicate packet information it hears to the APRS-IS internet network. When you do this, your computer will send the information.

4. To broadcast your location, you will need to configure the APRS program to utilize your GPS device. As you walk about, the radio that is equipped with APRS should be able to follow your position. Once, the location of the static station should be entered into the software's database.

5. To ensure that everything is functioning well, you should test your setup by sending and receiving APRS packets. If you are near a station that transmits APRS packets to the Internet, you may check a website like https://aprs.fi/ to see whether or not the packet was caught by the terrestrial network.

6. **Advanced**

- When designing a radio and computer, you may send your packets by connecting the output of your personal computer to the input of the radio. If you want to make your own "**Rigblaster**," you can either buy one or make one yourself.

How to Use APRS

After APRS has been configured, you will be able to begin utilizing it to communicate with one another and to monitor your whereabouts. It is possible to send APRS messages not just between stations but also to the APRS network, which may subsequently be shown on a map of the current location in real-time. It is common practice to make use of APRS to monitor the whereabouts of vehicles or individuals, such as search and rescue teams or hikers. Through the use of weather stations that are linked to the APRS network; you can use it to broadcast weather data, which includes temperature, pressure, and wind speed figures.

Advanced APRS Features

To improve its functioning, APRS is equipped with a wide variety of sophisticated features that can be exploited.

- Automatic position reporting and monitoring enables stations to automatically communicate their location at regular intervals. This feature helps track moving vehicles like ambulances and delivery trucks thanks to its ability to send such information.
- APRS's range may be increased by using a technique called "**digipeating**," which enables packets to be sent from one station to another.
- As a result of the efforts of the ARISS group, the International Space Station is equipped with an operational APRS digipeater. The frequency at which it runs is 145.825 MHz. PATH designators are one-of-a-kind, so make sure you verify them before you operate.

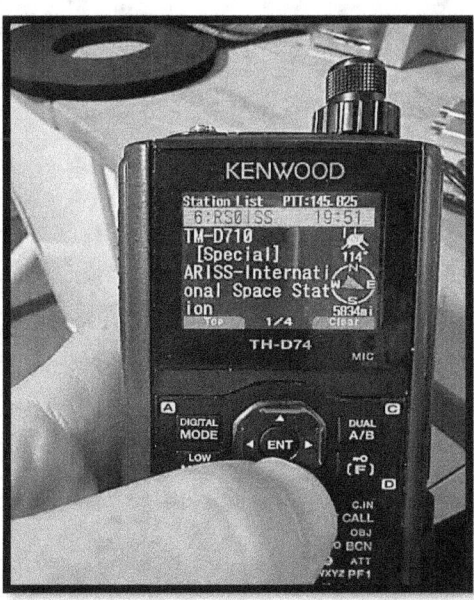

- APRS's ability to provide meteorological data, including temperature, humidity, and wind speed, is one of its many applications.

- Link weather stations to the APRS network to receive and report weather data.
- Aspiring hams participate in contests known as "**Golden Packets**," in which they try to deliver an APRS packet across vast distances using simplex transmission.

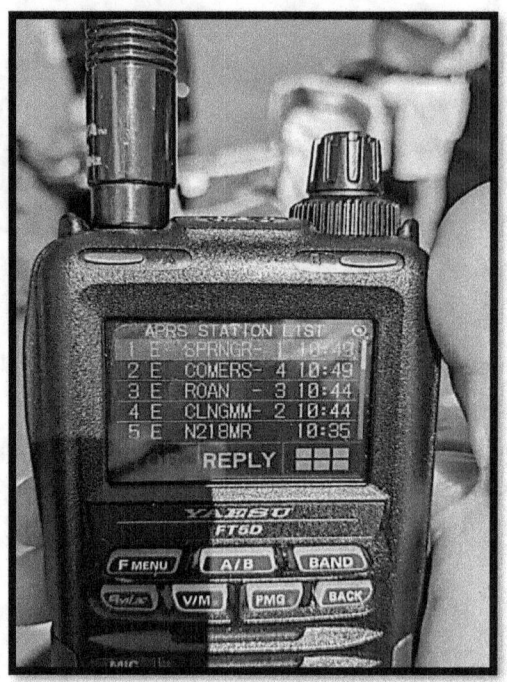

Best Practices for APRS

Following best practices is very necessary to run APRS efficiently.

1. To verify that your station is compatible with other APRS stations in your area, the first step is to check that it is set to the appropriate frequency and beacon rate.
2. It is of the utmost importance to establish appropriate pathways and filters to guarantee that your APRS packets are being sent to the appropriate stations.
3. Adhere to appropriate operating principles, which include avoiding excessive beaconing, avoiding the use of generic callsigns, and avoiding QRM.

Data Communication via Packet Radio

It was in the 1970s when ham radio gave birth to the concept of packet radio. It was first conceived as a method for transmitting digital data via the utilization of radio frequencies. Since it was first developed for use in radio applications, it has evolved into an essential component for the functioning of a variety of devices, including wireless routers, radio systems for local law enforcement and taxis, and analog cellular phone networks. An early effort to arbitrate and control many users over a shared communications channel was the focus of the experiments that were conducted with packet radio. By the middle of the 1970s, these efforts had developed into carrier sense multiple access (CSMA) protocols, which were the forerunners of EthernetTM configurations for local area networks (LANs). By 1978, the Federal Communications Commission (FCC) and other agencies had granted authority for ham radio operators to broadcast the American Standard Code for Information Interchange (ASCII) over radio transmissions. Early tests were conducted in the San Francisco region utilizing ham radio networks to validate packet data between mobile and stationary stations. These experiments were a result of this.

Direct sequence spread spectrum (DSSS) modulation and forward error control (FEC) technology were two examples of cutting-edge technology that was already in use at the time. These technologies were used to build data channels. These experiments included the Advanced Research Projects Agency Network (ARPANET), which was the forerunner of the Internet and was run by the United States Department of Defense. The ARPANET was responsible for the transmission of information between the packet radio network, satellite packet systems, and the ARPANET. This was done to facilitate an early version of internetworking. As a consequence, these advancements contributed to the development of some of the technology that underpins contemporary Internet Protocols (IP). For the ham operator, a typical early packet radio setup would include a keyboard, terminal, modem, transceiver with antenna, and terminal node controller. This would be the standard assortment of components. As part of its function, the computer would be responsible for managing network connections, formatting and packetizing data, and controlling the radio channel. Since the ham radio was intended for speech transmission rather than data transmission, technological challenges needed to be conquered. This rudimentary mix of the Internet and ham radio may potentially feature a simple message system for bulletin boards or provide other functionalities. A significant number of ham operators choose to use keyboard-based packet radio communication rather than speech transmission in the present day.

History of Packet Radio

The technology of data packets was created in the middle of the 1960s and was first put into practical use in the ARPANET, which was formed in 1969. With its beginnings in 1970, the ALOHANET, which was headquartered at the University of Hawaii, was the very first large-scale packet radio program. Montreal, Canada was the location where amateur packet radio was first broadcast, with the first transmission taking place on May 31st, 1978. After this, in 1980, the Vancouver Amateur Digital Communication Group (VADCG) developed a Terminal Node Controller (TNC). This was the next step in the process. A meeting of the Tucson Chapter of the IEEE Computer Society in October 1981 was the occasion for a debate that eventually led to the development of the current TNC standard. Another week later, six of the participants got together and spoke about whether or not it would be possible to create a TNC that would be accessible to amateurs at a price that was affordable to them.

Through the completion of this project, the Tucson Amateur Packet Radio Corporation (TAPR) was established. On June 26, 1982, Lyle Johnson, WA7GXD, and Den Connors, KD2S, were the ones who began a packet contact with the very first TAPR unit for the first time. Beginning with these prototype units, the project eventually evolved into the TNC-1, and then ultimately became the TNC-2, which serves as the foundation for the majority of packet operations around the globe.

Why Should You Choose Packet Over Other Modes?

Packet has three significant benefits over other modes: transparency, error correction, and automated control. The end user doesn't need to be aware of how a packet station works; all they need to do is connect to the other station, put in their message, and it will be sent automatically. An automated division of the message into packets is performed by the terminal Node Controller (TNC), which subsequently keys the transmitter and delivers the packets for transmission. During the process of receiving packets, the TNC does automated decoding, monitors for faults, and displays the messages that have been received. Error-free communications are made possible by packet radio thanks to the error detection systems that are included in the system. Once a packet is received, it is examined for any faults that may have occurred, and it will only be presented if it is accurate.

Additionally, any packet TNC may be used as a packet relay station, which is also referred to as a digipeater in certain instances. Through the linking together of many packet stations, this makes it possible to achieve a wider range. To determine whether or not their friends are at home, users can connect to their friends' TNCs whenever they like. Some TNCs also allow other amateurs to leave messages for them while they are not at home by providing them with Personal BBSs, which are also frequently referred to as mailboxes (for example, the Kantronics KPC-3 Plus). Another benefit of the packet over other modes is that it allows several users to utilize the same frequency channel at the same time. This functionality is not available with other forms.

Software-Defined Radios (SDR) Integration

From Analog to Digital: A Capsule History of SDR

When it comes to sending and receiving radio waves, analog electronics has always been the domain of responsibility. The fundamental operations of a radio, which include tuning, detecting, oscillating, mixing, filtering, (de)modulating, and amplifying, were accomplished via the use of crystals, capacitors, inductors, tubes, transistors, and other electronic components inside the device. To take advantage of the radio frequency (RF) spectrum, these components may be integrated into circuits that are tremendously sophisticated or even magnificent. Analog radios, on the other hand, are subject to constraints, the most significant of which is the extent to which they can function inside the spectrum. Any analog radio, except for the most complex ones, either has a frequency range that is quite limited or has various sets of components to enable operation in a variety of bands.

At the beginning of the 1970s, several research groups inside the United States government started experimenting with the possibility of replacing hardware components with software signal-processing methods. Math and code for digital signal processing (DSP) progressed at a quick pace at colleges and laboratories affiliated with the Defense Department. It became incredibly enticing to use software because it offered a method to circumvent many of the limits that were associated with hardware. The software had immense power and flexibility. In the 1990s, research and development of software radio, a phrase that would eventually be replaced by software-defined radio (SDR), quickly expanded out of the government and into commercial endeavors. By 1997, software radio was already being used in high-end automobile radio installations. As a result of the introduction of the European digital television standard known as Digital Video Broadcasting–Terrestrial (DVB-T) in 1997, manufacturers such as Realtek and others started producing low-cost integrated circuits such as the RTL2382U. These circuits were able to decode DVB-T signals in the range of 174MHz–786MHz and were packaged in small USB dongles. Between the years 2010 and 2012, Eric Fry and Antti Palosaari determined that the RTL2382 could be used as a general SDR.

How do SDRs Work?

In a demodulator chip like the RTL2382U, the analog-to-digital converter (ADC) is the core component of a single-diode receiver (SDR). This is the component that is responsible for converting the analog electrical signal that the radio antenna receives into a series of ones and zeros, in a manner that is analogous to the process of digitizing a signal from a microphone. To transmit the desired frequency range from the antenna to the demodulator, a digital tuner chip, such as the R820T, simply collects the frequency range from the antenna. The power of mathematics may be used via digital signal processing (DSP) software to filter, decode, and convert the data into targets of interest after the signal has been able to be represented numerically.

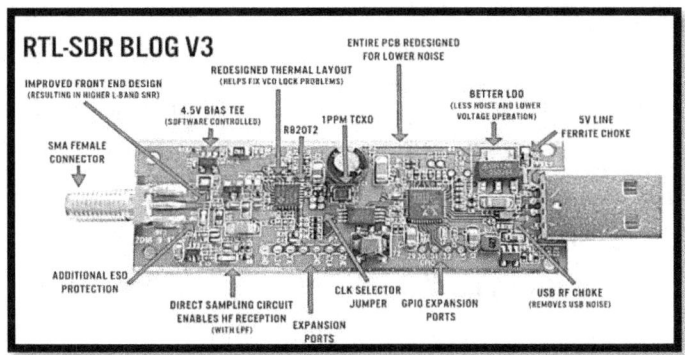

In actuality, real-world problems need the making of certain concessions. Examples of situations in which amplification is required include situations in which the analog signals may be extremely weak. For a variety of reasons, including filtering, noise, dynamic range, and others, amplifiers come with their own set of challenges. Bandpass filters are often used to restrain these issues; nevertheless, they restrict the array of frequencies that may be obtained. Multiple analog input channels are used in many SDR implementations. Each of these channels is equipped with optimal amplification and filters to provide consistent results throughout the widest possible range.

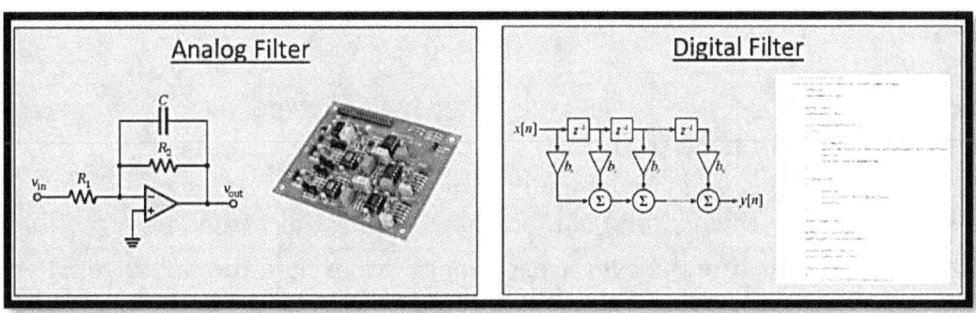

Since Bell Labs placed the first T-1 network into operation in 1962, digital signal processing (DSP) software has been the engine at the core of the tremendous technological advancements that have occurred simultaneously. It is possible to modify audio, video, seismological, medical, or any other signal with changes that can be monitored over time by shedding the constraints of physical attributes that limit the development of analog circuits. This allows for a degree of accuracy and flexibility that analog designers can only dream of.

Getting Started in SDR

An antenna, an SDR peripheral, and a computer are the three pieces of equipment that are required to begin exploring with scanning direct reception (SDR). In general, the RTL-SDR USB dongle, which can be purchased for around thirty dollars, is the most common starting setup for novice SDR software users. In addition, there is a large variety of more complex SDRs that may be purchased at a variety of pricing points. The frequency range of some of them is wider, they can transmit, and they have more sophisticated electronics. On the other hand, the little RTL-SDR is a fantastic place to begin.

Although many SDRs are capable of decoding data in the lowest end of the spectrum (which varies depending on the SDR and is often below 300MHz) by direct sampling or direct conversion, the clarity of these signals is sometimes diminished. As an alternative, many owners of SDRs may buy a cheap upconverter, which is a device that transforms signals that are below the SDR's reception range, often down to "**DC**" (0Hz), into frequencies that are within the standard range of communication for the SDR. Users who have more than casual listening requirements might benefit from the additional features offered by more complexes SDR systems. A receiver with powerful DSP filtering is available from Airspy in the sub-$200 price range. This kind of filtering is always useful for cutting through noise.

Antenna Considerations

Just as is the case with the majority of radio applications, the antenna is often the factor that contributes the most to the overall experience rather than the radio itself. Receiving antennas is more forgiving than sending antennas, although all antennas are tuned for certain frequency ranges.

SDR Software

It is common for many modern devices to have hardware that serves as the generalist and software that serves as the specialist. SDR software shows, filters, and decodes the broadband signal that is produced by the SDR device. It then converts the acquired data into signals that may be used.

The use of digital radio protocols has grown so widespread that, in the absence of software, a significant portion of the spectrum's fascinating activity would be unable to be understood. When it comes to mapping ADS-B position signals from aircraft, tracking frequency-hopping trunk systems, decoding digital FM broadcasts, operating as a spectrum analyzer, or recognizing satellites that are now above, SDR software is a wonderful example of how the usefulness of SDR hardware completely bursts. To find anything to listen to, the majority of people who are just starting with SDRs will scroll the waterfall display around the frequency range.

A next-level collection of activities may be found in the form of online tutorials that are designed to help individuals pursue specialized hobbies such as trunked police radio, aircraft tracking, or satellite telemetry. One would be hard-pressed to find a laptop, desktop computer, Raspberry Pi, tablet, or phone that cannot run some form of SDR software in this day and age. Operating systems such as Linux, Android, MacOS, and iOS all offer sophisticated capabilities, but Windows has the most extensive collection of applications. SDRangel and CubicSDR are two of the rare applications that are compatible with all three operating systems: Windows, Mac, and Linux.

Connecting and Operating SDRs with Ham Radios

Software Defined Radios, also known as SDRs, have gained a lot of popularity in the field of amateur radio, often known as ham radio, since they are versatile, inexpensive, and have the capacity to function over a broad variety of frequencies. The process of connecting and using SDRs with ham radios comprises a few critical phases, and this tutorial will offer a full description of the process along with the processes involved.

Understanding SDRs and Ham Radios

Software-defined radios, often known as SDRs, are radios that employ software to carry out a variety of operations that were previously carried out by hardware. The fact that these radios are capable of tuning over a large variety of frequencies and modes makes them very adaptable instruments for anyone interested in amateur radio. SDRs such as the RTL-SDR, HackRF, and Airspy are among the most popular.

The following are the steps to connect and operate SDRs with Ham Radios:

1. Choose an SDR that is suitable for your needs and preferences. Think about things like the frequency range, the modes that are supported, and the capabilities of the gear. For ham radio, the most common SDRs are RTL-SDR, which is used for receiving and HackRF or LimeSDR, which is used for developing more sophisticated capabilities.
2. The SDR should be connected to your computer using the proper interface, which is often a USB cable. Also, check to see that the appropriate drivers for your SDR hardware have been installed for your system.
3. Select SDR software that is compatible with the SDR gear you use. GQRX, HDSDR, and SDR# (SDRSharp) are some of the prominent choices available. Obtain the software of your choice and then proceed to install it on your computer.
4. When connecting the SDR to your ham radio transceiver, you should make use of an appropriate interface. This may need connecting the output of the SDR to the input/output ports of the transceiver. Make certain that the connections are safe, and make sure to adhere to any special instructions that are supplied by the makers of the SDR and ham radio components.
5. Start the software for the SDR and make sure it is configured to operate with the SDR gear you have. Then adjust the frequency, mode (AM, FM, SSB, etc.), and any other characteristics that are suitable for the purpose that you want to put them to.
6. Make sure that the SDR is calibrated so that it can receive frequencies accurately. This is of utmost significance for applications that are sensitive to differences in frequency.
7. Activate the software for the SDR, and keep an eye on the output. To see the signals that have been received, you should be able to view the spectrum, the

waterfall, or any other visual representations. The ham radio transceiver should be used for transmitting, and the SDR should be used for receiving. This configuration makes it possible to have a more all-encompassing picture of the spectrum.

8. You must get familiar with the rules and band plans that are relevant to amateur radio in your location. Make certain that the laws and regulations that have been established by the appropriate authorities are adhered to by your actions.

Frequently Asked Questions

1. What are the different types of digital modes in Ham Radio?
2. How do you explain data communication via packet radio?
3. What are the software tools for digital operation in Ham Radio?

CHAPTER SIX

EMERGENCY COMMUNICATIONS AND ARES

Overview

Ham Radio is often linked with emergency communication. This chapter talks about the role of hams in emergency communication, Skywarn & weather spotting, and community disaster preparedness.

ARES (Amateur Radio Emergency Service)

Role of Hams in Emergency Communication

Amateur radio operators, who are sometimes referred to as "hams," play an essential part in the transmission of emergencies and the response operations to disasters. Their participation becomes especially important in situations when the regular communication infrastructure is either disrupted or overwhelmed. A closer look at the function that hams play in the field of emergency communication is as follows:

Independent Communication Infrastructure

- **Redundancy**: Amateur radio can offer a communication network that is both redundant and independent. Traditional communication systems including landlines, mobile towers, and the internet may malfunction or become overloaded during emergencies. Ham radio operators can establish communication linkages even when other systems are compromised since they are equipped with their own transceivers and power sources.

Immediate Deployment

- **Rapid Response:** Hams can quickly deploy their equipment to impacted regions without having to depend on the infrastructure that is already in that location. Due to their capacity to build portable and temporary stations, they can facilitate the creation of communication in areas that have been affected by a catastrophe.

Flexible Communication Modes

Hams can communicate in a multitude of modes, including voice, Morse code, digital modes, and satellite communication. This gives them a high degree of versatility. Because of this flexibility, efficient communication may be maintained under a wide variety of circumstances, and it can be adapted to precisely meet the needs of the emergency scenario.

Communication in Extremely Remote Locations

It is possible to access distant or isolated locations with the help of amateur radio, which is a kind of communication that is successful in reaching areas where other means of contact would be impossible. This capacity is particularly essential in the event of natural catastrophes like earthquakes, hurricanes, or floods, which have the potential to cause significant damage to traditional infrastructure.

Providing Assistance to Emergency Services

Hams often work together with emergency services and first responders to ease communication and coordination efforts. This is one of the many ways that Hams contribute to the field. Critical information, such as medical needs, resource requirements, and situational updates, may be sent across the various reaction teams via their ability to convey this information.

Community Assistance

- **Public Service:** A significant number of amateur radio operators are actively engaged in activities that are considered to be public service, and they often do emergency drills and exercises. As a result of this proactive participation, partnerships with local emergency management agencies are strengthened, which in turn ensures a more smooth integration during real emergencies.

Network of Skilled Volunteers

Ham radio operators go through training and get licenses, which allow them to acquire technical skills that are useful assets during emergencies. They can diagnose and fix problems with communication equipment, which guarantees that their systems will continue to work normally.

Weather Monitoring and Early Warning

Skywarn: Hams are often engaged in initiatives like Skywarn, in which they offer early warnings for severe weather disasters and make contributions to the monitoring of meteorological conditions. The community's readiness is improved by this proactive strategy, which also has the potential to considerably lessen the effect of natural catastrophes.

International Collaboration

Global Reach: Due to the international character of amateur radio, it is possible to communicate with people all over the world. Hams can establish communication connections with operators in other countries in the case of a crisis, which makes it easier for them to share information, resources, and knowledge.

Engaging the Community and Providing Education

Education and preparation: Hams contribute to the preparation of the community by educating the general public about the role that amateur radio plays in response to emergencies. Training sessions and exercises are often held by them to improve the capabilities of operators as well as the broader civilian population.

ARES Organization and Deployment

Through the establishment of structured communications trunk lines and net systems, the Amateur Radio Emergency Service (ARES) was established by the National Traffic System (NTS) in conjunction with the Amateur Radio League (ARRL). Local Amateur Radio Emergency Service (ARES) organizations are a comprehensive and close volunteer group of radio amateurs and administrators who work together to create their own effective and different kinds of public-service communications. These groups operate as a component of the ARRL Field Organization.

The Amateur Radio Service (ARES) of Santa Cruz County is made up of radio operators who are licensed for the Amateur Service and have registered their credentials and equipment to conduct communications for the benefit of the general public. These kinds of communications are made available in the event of emergencies brought on by natural disasters (such as earthquakes and floods) as well as man-made disasters (such as chemical spills and other similar incidents). Our service area encompasses the whole of Santa Cruz County and consists of four ARES groups that work together but are operated independently. We also provide service to the areas of north Santa Cruz County and north Monterey County, which are located to our south. This is accomplished via collaboration with the ARES organizations that are located in their respective regions.

The Functions of the ARES

Part 97 of the Rules and Regulations of the Federal Communications Commission outlines the following as the basis and purpose of the Amateur Radio Service:

To provide emergency communications, "(a) Recognition and enhancement of the value of the amateur service to the public as a voluntary non-commercial communication service concerning the provision of emergency communications." ARES responds to requests for help that are made by agencies that are in charge of declaring emergencies (such as the County OES) or providing services during times of disaster. Every catastrophe is different, and there is no way to ensure that one is adequately prepared for the specifics of any upcoming crisis by preplanning. The best course of action for emergency groups is to be ready for as many different scenarios as they can. Both the planning and training that are carried out by ARES have as their major objective the preparation of the participants, as well as the ongoing connection with the agencies that we are responsible for.

You can get the assistance of ARES to provide communication services in situations where there are no established linkages or to enhance existing systems if they become dysfunctional or overburdened. It is the responsibility of public safety groups to ensure that their communication systems can appropriately manage emergency circumstances. To fulfill the peak demands that are caused by big catastrophes, it is not possible for them to consistently maintain the necessary resources. The purpose of radio amateurs in these situations is to supplement the communications capabilities that are already available to the government and disaster agencies. Members of ARES provide extra value to the services that we provide as a result of the additional capabilities that each of us has individually. For instance, our skills in troubleshooting and repair, our training and expertise with the Red Cross, our familiarity with public agency communications dispatch, and so on.

ARES offered services

The following are some examples of the services that Santa Cruz County Amateur Radio Emergency Services may offer:

- Communications between Santa Cruz County, other surrounding counties, and other Government agencies.
- Interactions between county authorities and other officials of local government or state agencies; such interactions may take place.
- Communications between county, municipal, and state public service agencies.
- A variety of disaster relief groups, such as the American Red Cross and the Salvation Army, get supplementary communication services.
- Supplemental communication services to hospitals and other medical resources.
- Health and Welfare messaging for the general population.
- Any further communications for the public service must be provided.

Training and Preparedness for Emergency Situations

Training and being ready for any kind of emergency circumstance are two of the most important parts of amateur radio, which is more frequently referred to as ham radio. When conventional communication lines may fail during emergencies, ham radio operators play a crucial role in providing communication assistance. In the field of ham radio, the following is an in-depth summary of the training and readiness procedures that are in place for emergencies:

Training

- The appropriate regulatory authorities should be contacted to get the proper ham radio license. The operators are granted access to a variety of frequency bands and modes according to the license level they possess.
- Acquire knowledge of core radio operating processes, such as proper protocol for communication, frequency use, and etiquette. You should have a fundamental understanding of the fundamentals of antenna systems, radio wave propagation, and electromagnetic waves.
- Familiarize yourself with the protocols and procedures that are used for emergency communication, such as the Amateur Radio Emergency Service (ARES) organization and the Incident Command System (ICS).
- Acquire advanced skills in the operation of a wide range of ham radio equipment, such as transceivers, antennas, power supplies, and accessories.
- Get knowledge of digital communication modes such as Packet Radio and Winlink, in addition to other ways of data transport. Your familiarity with the appropriate software for these modes should be emphasized.
- To become a certified weather spotter, you need to participate in training programs offered by Skywarn. In the case of severe weather, having this information is quite beneficial.
- To give prompt help during emergencies, you need to get certification in basic first aid and cardiopulmonary resuscitation (CPR).
- To prepare for potential emergency situations, you should practice establishing and running portable stations in a variety of field settings.

- Gaining expertise in managing large-scale communication operations and working with other agencies may be accomplished by taking part in public service activities.

Preparedness

- It is necessary to devise communication strategies for a variety of emergencies, such as those involving natural disasters, power outages, and public events. Check if the preparations are in line with the emergency management agencies in the area.
- Make sure that all of the radio equipment is in excellent operating order by doing routine inspections and maintenance on it. Ensure that you have sufficient backup power sources, extra batteries, and needed equipment at your disposal.
- Install and test a variety of antenna systems that are suited for a variety of frequencies and circumstances of propagation. Antennas that are lightweight and simple to deploy are very helpful in times of emergencies.
- The establishment of networks with local emergency management agencies, ARES organizations, and other amateur radio operators is another important step. Create connections that make it easier to coordinate effectively in times of emergencies.
- You are responsible for keeping an inventory of these resources, which should include radios, antennas, batteries, and any other necessary supplies. Ensure that products are regularly updated and replaced as required.
- It is recommended that emergency communication drills and exercises be carried out regularly to evaluate the efficiency of the emergency plans, as well as the equipment and communication protocols.
- Maintain an awareness of the current weather conditions and any developments. During times of emergencies, weather information is often relayed by Ham operators, who play an important role.
- The local people should be made aware of the function that ham radio operators play in the event of major emergencies. Individuals should be encouraged to earn their licenses to operate amateur radio.
- Maintain a level of awareness about emerging technology, laws, and the most effective methods of emergency communication by engaging in ongoing education and participating in training programs that are pertinent to the field.

- Ensure that comprehensive documentation of communication strategies, specific configurations of equipment, and contact information for important stakeholders is maintained. When conditions are very stressful, this record is of the utmost importance.

Skywarn and Weather Spotting

Collaborating with the National Weather Service (NWS)

The Skywarn program is a program run by the National Weather Service (NWS) that is comprised of trained volunteers who identify hazardous weather. Through the provision of timely and accurate severe weather reports to the National Weather Service (NWS), Skywarn volunteers assist their local community and government.

The integration of these reports with contemporary NWS technology allows for the dissemination of information to communities on the appropriate steps to take if severe weather is imminent. The National Weather Service (NWS) has relied on Skywarn, which was established in the early 1970s, to offer vital information on severe weather promptly, allowing for the required warnings to be delivered. As a result, the primary objective of the Skywarn program is to preserve lives and property by making use of the observations and reports received from volunteers who have received training. (2003) Gropper (1993). The National Weather Service is only able to assess the likelihood of severe weather, despite the sophisticated radar and forecasting technologies that they have at their disposal. Reports from members of the public and law enforcement professionals, as well as real severe weather, are what they depend on. It might be challenging to get information from the general population that is both accurate and of high reliability.

Extensive weather is a tricky and perplexing phenomenon. Only through consistent training of weather spotters can the National Weather Service (NWS) increase the quality of the information they provide. To coordinate the development of training programs, the National Weather Service (NWS) works in conjunction with Amateur Radio clubs and other organizations. Together with the local government and the Red Cross, the National Weather Service (NWS) contributes its experience in weather forecasting, and the Amateur Radio Service (ARS) contributes its understanding of emergency communication technology.

A Memorandum of Understanding (MOU) between the Amateur Radio Relay League (ARRL) and the National Weather Service (NWS) publicly acknowledges and encourages the involvement of Amateur Radio operators in the Skywarn program. This agreement indicates that the American Radio Relay League (ARRL) would encourage its local volunteer organizations that operate as the Amateur Radio Emergency Services (ARES) to provide spotters and communicators to the National Weather Service (NWS) in response to requests made by the NWS during times of severe weather. (2003) Gropper (1993). There are a great number of civic catastrophes that are either directly caused by severe weather or are made worse by severe weather. As a result, the National Weather Service (NWS) may make use of the Skywarn Amateur Radio operators not only to acquire and disseminate severe weather observations and warnings but also to maintain close coordination with the Red Cross and Emergency Managers from local government entities that fall under the umbrella of ARES or Radio Amateur Civil Emergency Service (RACES) (Gropper, 1993). RACES are an organization that is comprised of free-lance Amateur Radio operators who have received training in emergency communications and the detection of severe weather. In times of emergencies, the Regional Emergency Communications and Emergency System (RACES), which is authorized and managed by the Federal Emergency Management Agency (FEMA), offers vital communications and warning linkages? During the aftermath of Hurricane Andrew in August of 1992, Skywarn was able to highlight the significance of this extra function that they play.

Skywarn observers who have received training supply the Weather Service with report information that is both accurate and timely from radio-equipped vehicles and houses. Extreme weather reports are the ones that the NWS is most interested in. The following are examples of severe weather: flash floods, hail, destructive winds, a wall cloud (which is the region of a thunderstorm where a tornado might occur), and a tornado funnel. The National Weather Service (NWS) will notify local authorities, who will then be able to activate Civil Defense sirens if it determines that severe weather is occurring using radar and other available information. Notification is sent to the news media so that they may report on the incidents that occur on local broadcast stations. To provide their communities with early notice of potentially life-threatening weather, Skywarn volunteers generously offer hundreds of hours of their time and use their radio equipment and vehicles. The number of fatalities that have occurred as a result of tornadoes and other forms of severe weather has significantly decreased ever since the National Weather Service (NWS) launched the Skywarn Program.

To provide a comprehensive perspective of the partnership between ham radio operators and the National Weather Service, the following is provided:

Purpose of Collaboration

- The major objective of this partnership is to improve public safety by delivering accurate and timely weather information to the community. This will be accomplished via the provision of weather information.
- The gathering of data Ham radio operators can contribute to the collection of real-time weather data, which assists the National Weather Service in improving its performance in predicting.

SKYWARN Program

- The National Weather Service (NWS) and ham radio operators can work together more effectively thanks to the SKYWARN program.
- Volunteers who work with SKYWARN get training that enables them to see and report severe weather occurrences, including but not limited to thunderstorms, tornadoes, hail, and other meteorological experiences.

Responsibilities and Roles to Consider

- **Ham Radio Operators**: Ham radio operators serve as the eyes and ears concerning the situation on the ground. They keep an eye on the weather and send their observations to the National Weather Service (NWS), which provides vital information for weather monitoring and forecasting.

- The National Weather Service (NWS) meteorologists are responsible for receiving and analyzing the reports that are sent by ham radio operators. With the use of this information, they can provide timely weather alerts and warnings to the general public.

Training and Certification

- The National Weather Service (NWS) or local emergency management agencies are often the organizations that give training to ham radio operators who are interested in weather reporting.
- Techniques for observing the weather, report writing, and the fundamentals of recognizing severe weather are covered in the training.

Communication Protocols

- To guarantee that the information that is sent between ham radio operators and the NWS is handled in an effective and precise manner, standardized communication protocols are devised.
- During times of high stress, these guidelines serve to speed the process of reporting to the appropriate authorities.

Integrated Warning System

- As part of the partnership, ham radio communication is often included in the larger emergency alert and warning system.
- The overall efficiency of the warning system is improved as a result of this measure, which guarantees that information about severe weather is disseminated to the general public via a variety of channels.

Amateur Radio Emergency Service (ARES)

- There is a significant contribution that ARES organizations, which are often associated with the American Radio Relay League (ARRL), make to the field of emergency communications. To offer communication assistance during severe weather occurrences, they work along with the National Weather Service.

Technological Integration

- Recent technological advancements have made it possible to collaborate more seamlessly. The effectiveness of information exchange between ham radio operators and the NWS is improved by automated weather stations, mobile applications, and digital communication modalities.

Identifying and Reporting Severe Weather Conditions

One of the most important aspects of ham radio operations is the identification and reporting of particularly severe weather conditions. Ham radio operators, who are also referred to as amateur radio operators, play a significant role in the provision of information about weather conditions that are both real-time and on the ground during and after emergencies. They serve as an essential link between the affected areas and the agencies that respond to emergency situations.

Identifying Severe Weather Conditions

Weather Monitoring

- Maintain a consistent monitoring schedule for weather reports obtained from official sources, such as the National Weather Service (NWS), meteorological agencies, or any other authorities that are pertinent.
- Ensure that you are up to date on the most recent weather information, particularly during severe weather events such as hurricanes, tornadoes, floods, and severe storms.

Understanding the Weather Patterns

- Familiarize yourself with the various types of severe weather conditions, including their characteristics, patterns, and the potential impacts that they could have.
- Recognize important weather indicators, such as shifts in the pressure of the atmosphere, patterns of wind, cloud formations, and temperature variations.

Use Weather Instruments

- Ham radio operators frequently have access to a variety of weather instruments, such as thermometers, barometers, and anemometers. To monitor and measure the weather conditions, you should learn how to use these instruments.

Reporting Severe Weather Conditions

Follow the steps below:

- During extremely severe weather events, ham radio operators have the opportunity to participate in emergency nets. Nets are structured communication channels that allow operators to share information in real-time and coordinate their responses.
- Ensure that you follow the procedures that have been established for reporting severe weather conditions. To effectively transmit information, this may involve the use of particular frequencies, codes, or formats.
- When reporting severe weather conditions, it is important to include essential information such as the sort of weather event that occurred, the location, the time, and any damage or impacts that were observed. Make sure that your communication is both clear and concise.
- Collaborate closely with relevant authorities, such as the National Weather Service (NWS) or local emergency management agencies. It is important to provide them with information that is both timely and accurate so that they can make more informed decisions.
- During times of emergencies, certain amateur radio organizations have established nets for severe weather. You should tune in to these nets and follow the procedures that they have established for reporting and coordinating information.

- To report severe weather conditions, you should put your safety first. It is important to avoid putting yourself in danger and to report information from a relatively secure location.

Community Disaster Preparedness

Ham Radio's Role in Community Resilience

Emergency Communication:

- Ham radios can function independently of conventional communication infrastructure, such as cell towers or the Internet. This is achieved by their independence from the infrastructure. During times of crisis, when normal communication lines may be disrupted, this independence is of the utmost importance.
- Ham radios often offer superior coverage and dependability than other communication techniques, particularly in locations that are rural or affected by natural disasters, when other sources of communication may fail.

Rapid Deployment:

- Ham radio equipment is intended to be portable and can be rapidly set up in a variety of settings. Establishing connections in regions where infrastructure has been damaged or destroyed is made much easier by the presence of this functionality.
- Ham radio operators often serve as members of emergency response teams and can quickly deploy their equipment to offer communication assistance both during and after catastrophes.

Community Coordination:

- When it comes to local networks, Ham radio operators are often deeply ingrained in the areas in which they operate. They can immediately build communication networks, which make it easier for members of the community, emergency services, and relief groups to coordinate their efforts simultaneously.
- Hams can communicate vital information like evacuation routes, emergency shelters, and resource distribution. This ensures that people of the community are kept informed and that choices are made based on current information.

Public Service:

- Ham radio operators are individuals who are actively engaged in public service and are considered to be volunteers. They take part in activities such as marathons, parades, and public meetings, where they provide assistance with communication and watch out for the participants' safety.
- Ham operators can assist in search and rescue operations during emergencies by effectively communicating information amongst various teams and coordinating their efforts.

Training and Education:
- Ham radio clubs often provide training sessions and drills, which assist members of the community in becoming acquainted with the protocols for emergency communication. As a result, the overall preparation and resilience of the community is improved.
- Obtaining an amateur radio license entails studying radio operations, rules, and emergency protocols. This information is transmitted across the ham radio community and contributes to a better educated and prepared people.

Global Connectivity:
- Ham radio provides worldwide communication. In the aftermath of calamities, ham operators throughout the globe may unite to give aid, information, and support to impacted people, transcending national lines.

Technology Innovation:
- Ham radio always advances with technology. The incorporation of digital modes, satellite communication, and other developments boosts the capability of ham radio in varied scenarios, leading to more effective communication tactics during emergencies.

Collaborating with Local Emergency Services

Collaborating with local emergency services via ham radio means creating efficient communication channels to help with emergency response activities. Amateur radio operators, or **"hams,"** play a key role in providing communication services when conventional networks may fail during emergencies.

- Familiarize yourself with the emergency communication protocols and procedures established by local emergency services. This includes learning how they cooperate, the frequencies they use, and the line of command during emergencies.

- Ensure that you and your fellow ham operators have the required licenses to operate on designated emergency frequencies. Work with local officials to secure any required licenses and approvals.
- Build ties with local emergency service workers, such as police, fire, and emergency management officials. Attend neighborhood meetings, training sessions, or other activities where you may introduce yourself and discuss the potential of ham radio in emergency circumstances.
- Actively engage in emergency drills and exercises offered by local emergency services. This enables you to practice communication techniques, identify possible obstacles, and show the advantages of ham radio in a controlled setting.
- Collaborate with local emergency agencies to establish specialized communication requirements for various sorts of emergencies. Understanding these demands can help you design your ham radio capabilities to deliver the most effective service.
- Work with local authorities to identify specialized frequencies for emergency communication. Ensure that all ham operators engaged are aware of these frequencies and are taught to utilize them correctly.
- Designate places for emergency communication centers where ham operators may congregate and form a foundation for coordinating communication activities. These centers should be equipped with the required radio equipment, power supplies, and backup systems.
- Offer training sessions for local emergency service workers on fundamental ham radio operations. This might include how to operate ham radios, emergency communication procedures, and collaborative techniques during crisis scenarios.
- Regularly maintain and test your ham radio equipment to ensure it is in excellent operating order. Also, put contingency plans in place for power outages or other disturbances to provide continued communication assistance.
- A tight working relationship with the local emergency services should be maintained during genuine emergencies. Ensure that established processes are adhered to, that timely updates are provided, and that assistance is provided in the transmission of vital information across various agencies and places.

- Following each emergency or exercise, you should chronicle your experiences and then suggest areas in which you may improve. An evaluation of the efficiency of the communication methods and protocols should be carried out, and input should be shared with the local emergency services to improve future cooperation.
- You should become involved with the community to increase awareness about the function that ham radio plays in times of emergencies. It is important to encourage folks to get licenses to operate amateur radios and to engage in community emergency response measures.

Building and Maintaining Emergency Communication Kits

When it comes to ham radio, the process of constructing and maintaining emergency communication kits requires careful planning, the selection of components, and continuing maintenance to guarantee dependable functioning at times of greatest need. During emergencies, when conventional means of communication may be ineffective, amateur radio, or ham radio, is a crucial resource.

Building Kits for Communication in Case of Emergencies:

Identify your Communication Needs

- Identify the reason why you have assembled your emergency communication pack. Are you looking for communication on a local, regional, or long-distance level?
- Take into consideration the many kinds of emergencies that might occur in your region, such as natural catastrophes, power outages, and so on.

Choosing the Appropriate Radio Equipment

Portable Transceiver (HT)

- Make sure you get a trustworthy handheld transceiver (HT) that has enough power output.
- If you want greater versatility, you might think about dual-band capabilities.

Mobile/Base Station

- To get a greater range, it is recommended that you use a mobile or base station transceiver that has a larger power output.
- Ensure that the antennas are compatible with external antennas.

Antennas

- Antennas should be selected depending on the kind of transceiver and your communication demands.
- Portable antennas for base stations and mobile devices, as well as more powerful external antennas for mobile equipment.
- When it comes to rapid deployment, portable mast solutions should be considered.

Power Sources

- You should include solar chargers, rechargeable batteries, and backup battery packs in your package.
- Ensure that your base station is powered by a dependable source of energy, such as a generator, solar power, or a backup battery.

Generator for Power in an Emergency

- When dealing with extended emergencies, you might think about purchasing a portable generator.
- Make sure that the needs for fuel and maintenance are not overlooked.

Accessories

- For improved clarity of speech, headphones, microphones, and external speakers are all recommended.
- Make sure you have extra adapters, connectors, and cables on you.

Weatherproof and Portable Containers

- Ensure that the equipment is stored in waterproof enclosures.
- Take into consideration portability to facilitate a speedy evacuation if it is required.

Maintaining Kits for Communication in Case of Emergencies

They include the following:

1. To guarantee that the equipment is functioning properly, you should do routine radio inspections. Conduct regular inspections of the antennas and power supplies.
2. Always make sure you are up to speed on the latest firmware and software upgrades for your radio equipment. Ensure to update whenever it is required to guarantee compatibility and enhance performance.
3. It is important to preserve the lifetime of rechargeable batteries by charging and discharging them regularly.
4. The antennas should be inspected for any indications of damage or wear. To avoid rust, ensure that connections are clean.
5. Your emergency paperwork should be updated regularly. Ensure that new frequencies or connections are added.
6. Maintain a regular schedule of emergency communication drills training. Ensure that you are up to date on any newly implemented emergency protocols and procedures.
7. Take part in ham radio nets and activities that take place in your area. Establish a community of operators to assist one another.
8. Maintain a level of awareness about the local weather conditions and any emergency alerts. You should always be ready to activate your emergency communication equipment in case it is required.

Frequently Asked Questions

1. What are the roles of Hams in emergency communication?
2. How do you identify and report severe weather conditions?
3. How do you train and prepare for emergency situations using Ham Radio?
4. How do you build and maintain emergency communication kits?

CHAPTER SEVEN
ANTENNA DESIGN AND OPTIMIZATION

Overview

Chapter Seven fully discusses antenna design and optimization in amateur radios. Learn the various types of antennas and the functions they serve.

Antenna Theory and Fundamentals

Types of Antennas and their Characteristics

The antennas used for Ham radio exist in a wide variety of kinds, each of which has its own set of features, benefits, and drawbacks. Because of the wide range of features, sizes, and price points, rookie ham radio operators sometimes feel overwhelmed by the alternatives that are available to them. This is a very understandable reaction. Your antenna needs to be appropriate for the sort of ham radio you wish to use as well as the communication style you desire to use. Because of this, a mobile antenna is necessary for a portable transceiver that is designed to be used during disaster relief operations. On the other hand, a base station has the capability of accommodating a bigger and more complicated antenna. Since the majority of these antennas are categorized as high-frequency (HF) antennas, they are perfect for long-distance direct-to-ear (DX) transmission.

Antenna Wave Lengths

Before we go into the various kinds of antennas, let's take the time to answer a question that is often asked: Does the size of a ham radio antenna make a difference? There is no guarantee that a longer wavelength would improve performance; nonetheless, a larger is not always better. Based on the proportion of the wavelength of the radio signal, antennas are categorized into several categories. Quarter-wave antennas, half-wave antennas, and five-eighths-wave antennas are the most prevalent types of antenna options. A quarter of a wavelength is the length of a quarter wave antenna. It is usual practice to make use of the 1/4 wave antenna because it often offers the most optimal mix of antenna transmission and reception efficiency in situations when a smaller antenna size is particularly significant. The size of a 1/2 wave antenna is twice as large as that of a 1/4 wave antenna, and it may be preferable due to the possibility of a low impedance match between the antenna, the coaxial cable transmission line, and the receiver. The 5/8 is an excellent choice for FM broadcasting and communication over great distances.

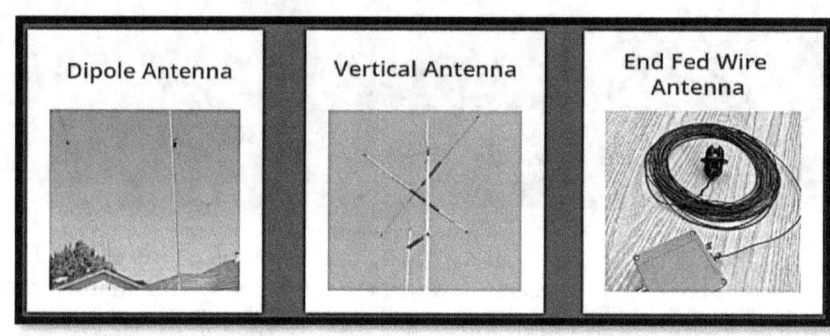

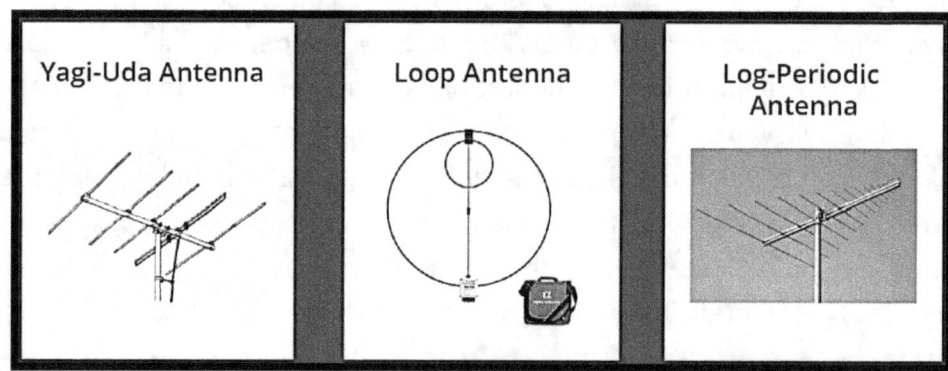

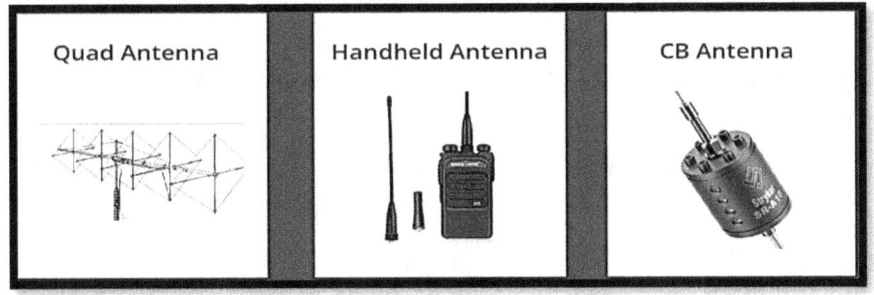

| Quad Antenna | Handheld Antenna | CB Antenna |

Dipole Antenna

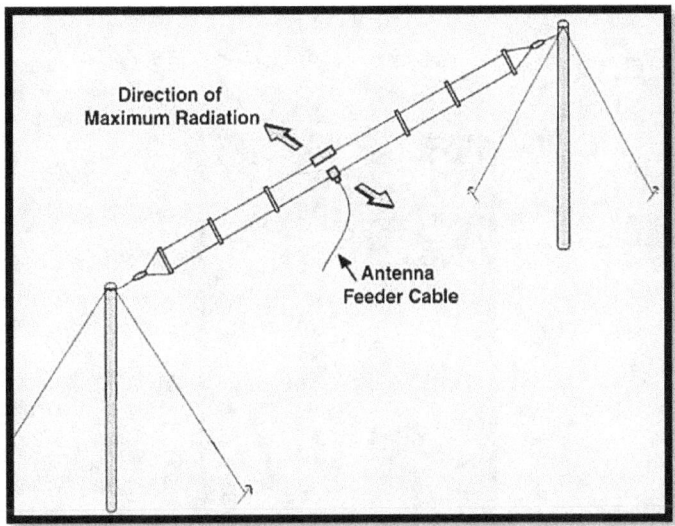

A dipole antenna, also known as a die-pole antenna, is often considered to be the most prevalent form of antenna system used by amateur radios. Dipole antennas are readily accessible in shops and online, and they are inexpensive. Additionally, they are easy to put up. A dipole antenna is made up of two metal rods or wires of equal length that are positioned end to end with a space in the middle where the feed line joins. The most common form of feedline that is used for a dipole antenna is coax cable; however, other types of feedlines, such as window lines or ladder lines, are also appropriate. In most cases, the dipole is almost half the length of a wavelength, which enables it to almost resonate on a single frequency. As a result of its efficiency, less voltage is required for transmission, and similarly, less amplification is required for receiving. Dual-band antennas make up the majority of wire dipole antennas now.

Pros

- The construction process is straightforward and uncomplicated.
- It is efficient in both sending and receiving.
- Capable of conveying information at a medium-range and local level

Negatives

- Limited bandwidth.
- For best performance, it is necessary to have the appropriate height and orientation.
- Easily susceptible to influence from items nearby

Vertical Antenna

An omnidirectional antenna is a form of antenna that transmits in all horizontal directions, making it useful for widespread communication. A vertical antenna is specifically designed for this purpose. A single radiating element is positioned vertically (upright) with radials to act as a ground plane. This is the only component that makes up this device. Gain potential increases in proportion to the height of the antenna; nevertheless, as the transmission line lengthens, there is a possibility that SWR will be sacrificed. It is possible to add a transfer device to get the impedance down to the benchmark of 50 ohms. These radials are made up of materials that are electrically conductive and serve as a ground plane due to their composition. Without them (or another element of a comparable kind), the vertical antenna will not function in a manner that is suitable for usage. Radials typically have a wavelength of one-fourth of a wavelength; however, extended wavelengths are possible if more materials and space are available.

Pros

- Omnidirectional
- It is straightforward to install and takes up little space.
- For long-distance communication, a low radiation angle may be advantageous.
- Can be constructed in a vertical configuration with a wavelength of 1/4, 1/2, or 5/8 for mobile applications.

Negatives

- There is a lower gain compared to directional antennas.
- Issues with vertical antennas are caused by noise that is caused by humans.
- The device is prone to experiencing poor performance on lower frequency bands and when it is transmitted over a short distance.

End Fed Wire Antenna

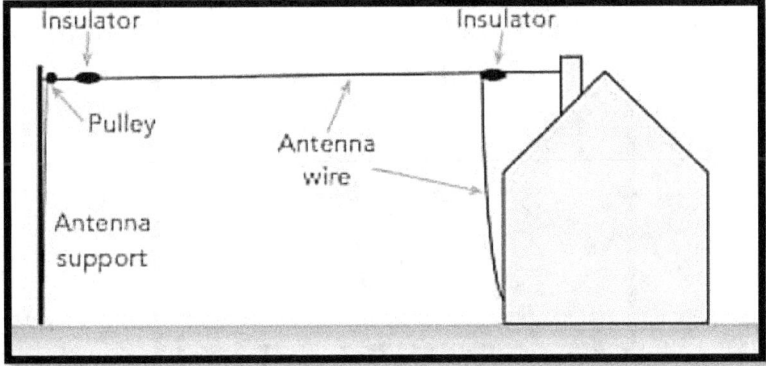

There are a few different varieties of end-fed antennas, including the end-fed random wire antenna and the end-fed half-wave antenna. The end-fed antenna is also sometimes referred to as a long-wire antenna. Even though end-fed antennas are still in use today, the long-wire variant of the antenna achieved its pinnacle of popularity in the middle of the 20th century. A significant benefit of end-fed antennas is that they are simple to install. Grounding these sorts of antennas may help reduce the amount of interference that they experience, which is a common problem. Matching the feedpoint to the transceiver may also be accomplished by operators via the use of an antenna tuner or tuning unit (ATU).

Advantages

- The cost is low.
- Lightweight and portable
- It is simple to set up.
- The capability to perform operations on many bands.

Negatives

- Noisy
- There is a high likelihood of interference from power lines and other electrical equipment.
- Interference with televisions, radios, and internet transmissions is another potential occurrence.

Yagi-Uda Antenna

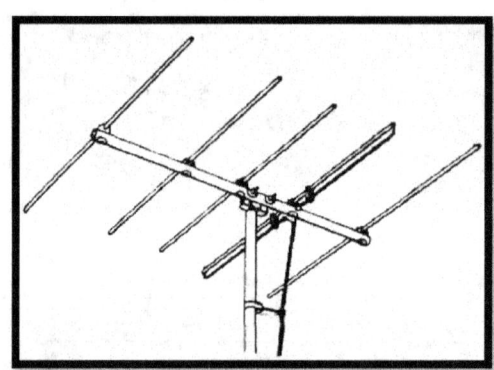

A driving element, a reflector, and one or more directors are the components that make up a Yagi-Uda antenna, which is sometimes referred to as a Yagi. This kind of antenna has directional characteristics. It is an antenna that only operates on a single band. The element that is being pushed is typically a dipole with a half-wavelength. "**Reflecting**" the radio waves back to the dipole is the function of the reflector, which is positioned behind the dipole. To increase the gain of the radio waves, directors serve the purpose of focusing on them. In general, the performance of the antenna will improve in proportion to the number of directors that are present. Typically, each of these components is installed on a horizontal boom and is arranged in a parallel fashion. In terrestrial communication as well as satellites, Yagi antennas are becoming prevalent. In addition to that, it is acknowledged that they are used in FM radio, television, and wireless networking.

Pros

- The signal intensity is enhanced due to the high gain and directivity.
- It is suitable for communication across vast distances.
- It is efficient in both sending and receiving.

Negatives

- Large and hefty, necessitating the establishment of appropriate support structures
- Aiming and placement must be done with great precision.
- Limited bandwidth.

Loop Antenna

A loop antenna is a form of antenna that is often supplied from the bottom and comprises a wire or metal loop inside its structure. It has the appearance of a steering wheel that is far larger than normal. There are two types of loop antennas: very tiny magnetic loops and very large resonant loops. On the other hand, electric loop antennas are used across VHF/UHF bands, which range from 30 MHz to 3 GHz, whilst magnetic loop antennas are commonly utilized for high-frequency (HF) communications.

These loop antennas are more compact than standard ham radio antennas, and they may be put inconspicuously on a rooftop or a window. Another advantage is that they can be used inside. While you are away on your next trip, you may use your magnetic loop to operate from your hotel room or RV. Due to the issues over their performance, loop antennas often have a negative reputation among those who utilize amateur radio in general. On the other hand, provided that they are installed correctly and in a suitable area, they are capable of performing effectively.

Pros

- The size is compact and tiny.
- Magnetic loops are possible to be transported.
- It provides a reception with a minimal level of noise.

Negatives

- Limited bandwidth.
- In the case of magnetic loops, the radiation efficiency is low.
- There is a need for precision tuning, which may be a challenging task for amateur radio enthusiasts who are just starting.

Log-Periodic Antenna

Another type of directional antenna that has a broad frequency range is called a log-periodic antenna. It is made up of several dipoles that are half the wavelength, with each one being cut for a certain frequency. A transmission line is used to provide it with power, and it is positioned on a boom. The strength of log-periodic antennas is such that they may be used over the whole spectrum, including sideband, FM, repeaters, moonbounce, and other frequencies.

Pros

- It has a broad frequency range.
- A high level of gain and directivity
- Suitable for several bands

Negatives

- Large and intricate
- There is a need for appropriate support structures.
- There is a possibility of its performance being diminished at the lower end of its frequency range.

Quad Antenna

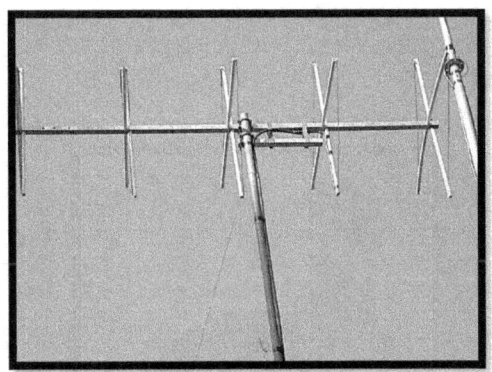

The term "**quad antenna**" refers to a directional antenna that is composed of one or more square or rectangular loops, each of which functions as an element (thus the name "**quad**"). Other names for this type of antenna include a cube antenna or a cubical quad antenna. It uses a whole wave rather than two-quarter waves in its operation. It can function on a variety of frequency bands, including UHF, VHF, and HF, and it may be programmed to operate as a single loop (mono-band) or many loops (multi-band). In quad antennas, the directional gain is proportional to the number of loops that are present. It is common practice to use copper tubing for the loop parts that are responsible for creating the square form. The spreader arms that are linked to the boom are what provide support for these loop parts. Since it is one of the most difficult antennas to construct, the calculations need to be accurate. They offer some of the greatest gains among all kinds of ham antennas, particularly when compared to the typical dipole antenna, even though they are not the most robust.

Pros

- Highly effective gain and directivity
- In terms of both sending and receiving, the performance is satisfactory.
- When compared to other antennas, it is less susceptible to being impacted by things that are nearby.

Negatives

- Bulky and needs correct support structures to be in place.
- Aiming and placement must be done with great precision.
- This antenna is more difficult to construct and tune in comparison to other antennas.

Handheld Antennas

You most likely have a default HT whip antenna with your radio unit if you are using a mobile ham radio. This antenna was included in the package itself. In most cases, these whip antennas, which are also known as rubber duck antennas, are quite straightforward. To connect to the radio unit, they consist of little more than a coil of wires that is encased in a small tube made of rubber or plastic and has a male/female SMA connection.

Although these portable antennas are suitable for use with the majority of individuals and circumstances, their range and clarity are restricted since they are somewhat short. To get more range, the majority of people who use portable ham radios switch out their normal antennas for either a whip antenna with a quarter wavelengths or a telescopic antenna.

Pros

- The portable ham radio is often combined with a simple and inexpensive device that is easy to operate.
- Extremely adaptable and versatile.

Negatives

- Cheaply built; they won't last long with frequent usage
- Due to its shorter length, the range is restricted.

CB Antenna

You may be surprised to learn that you can use a CB antenna with your 10-meter ham radio. This may seem like an exaggeration, but it is true. You read that correctly; if you come from the world of CB radio, you can use your preferred antenna on the 10-meter ham band (given that you make some adjustments to it).

Pros

- There are CB antennas that are simple to install since some of them just attach to the car.
- A satisfactory communication range is achieved.

- Truckers can use their CB antenna to operate a 10-meter radio.

Negatives

- You must adjust them by using an SWR meter.
- The modification of CB antennas may be necessary to either enhance or reduce their length.

Next Steps

Various types of antennas are designed to do a variety of tasks and are most effective in certain circumstances. There are several considerations to take into account while selecting a ham radio antenna, including the frequency range, the required communication range, the available space, and the budget. After gaining a grasp of a few of the most common high-frequency antennas, you are now in a position to go to the following phase, which is to investigate how each alternative meets your requirements.

Constructing Simple and Effective Antennas

A typical and satisfying component of amateur radio (ham radio) is the construction of simple and effective antennas. Antennas are essential components of any radio communication system, and designing and building your own may be informative as well as cost-effective.

Understanding Antenna Fundamentals

Frequency and Wavelength

- The first stage in creating an antenna is deciding which frequency or frequencies you wish to use. Each frequency has its wavelength, and the size of the antenna is proportional to the wavelength of the desired frequency.

How to Make a Simple Dipole Antenna

Materials Required

- Coaxial cable (RG-58 or equivalent)
- Insulators
- Balun (optional)

- Wire for the antenna elements

Steps:

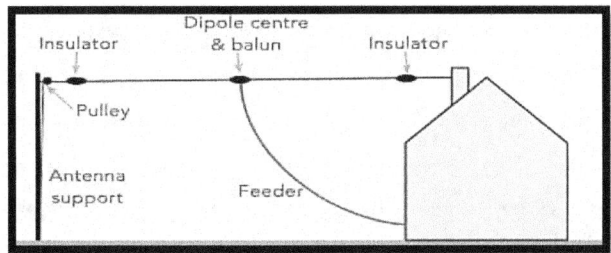

1. Select the frequency or band on which you wish to operate. A dipole is generally half the length of a wave.
2. Use the following formula: Length (in feet) =468 Frequency (in MHz) 468. Divide this length by two to obtain the length of each dipole side.
3. Cut the wire to the desired length. Make sure the wire is straight and taut.
4. Wrap insulators around each end of the wire. Plastic, ceramic, or any non-conductive material may be used to make them.
5. Connect the coaxial cable's center conductor to one side of the dipole and the shield to the other.
6. Raise or hang the dipole as high as feasible. It might be flat, slanted, or inverted V-shaped.
7. If you are using a coaxial cable, you might consider installing a balun (balanced to unbalanced transformer) to boost the performance of the antenna.

Understanding Yagi Antennas

Yagi Antenna

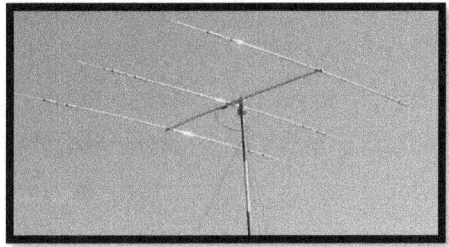

Yagi antenna theory

Yagi's theory requires a solid grasp of the phases of the currents flowing through the antenna's various parts. The Yagi antenna's parasitic elements work by re-radiating their signals in a slightly different phase than the driving element. As a result, the signal is strengthened in certain directions while being canceled out in others. Since the other antenna elements in the Yagi are not directly driven but instead receive power from the driven element, they are referred to as parasitic elements. One disadvantage of the Yagi antenna design is that the power in these extra parts is not directly driven. As a result, the produced current's amplitude and phase cannot be regulated. It is determined by their length as well as the distance between them and the dipole or driven element.

This indicates that total cancellation in one direction is not achievable. Nonetheless, it is still possible to have a high degree of reinforcement in one direction and a high level of gain, as well as a high degree of cancellation in the other to offer a favorable front-to-back ratio. The Yagi antenna may produce highly useful gain and front-to-back ratios.

An element can be made inductive or capacitive to achieve the needed phase shift. Each form of reactance has a distinct impact.

- **Inductive**: When the parasitic element is made inductive, the induced currents are discovered to be in such a phase that they reflect power away from the parasitic element. As a result, the RF antenna emits greater power in the opposite direction of the parasitic element. A reflector is an element that achieves this. By adjusting it below resonance, the element may be turned inductive. This may be accomplished by physically increasing inductance in the form of a coil, or by making it longer than the resonant length. It is often manufactured around 5% longer than the driving element to save money and retain the element mechanically as one piece, making it cheaper and stronger. A single reflector is always used. The installation of more reflectors makes no discernible change.
- **Capacitive**: If the parasitic element is capacitive, the induced currents will be in such a phase that they will direct the power emitted by the whole antenna in the direction of the parasitic element. A director is an element that performs this. It may be capacitively tuned above resonance. This may be accomplished by physically adding capacitance to the element in the form of a capacitor, or by

making it roughly 5% shorter than the driving element. It has been discovered that adding more directors boosts the antenna's directivity, boosting the gain and decreasing the beamwidth. The duration of subsequent directors is gradually shortened.

Multi-Band and Tunable Antenna Options

Ham radio communication depends heavily on antennas. They are in charge of transmitting and receiving radio signals, and the antenna used may have a considerable influence on the functioning of a ham radio station. Since they provide versatility and efficiency across several frequency bands, multi-band, and tunable antennas are popular among ham radio operators.

Let's take a closer look at these antenna options:

Multi-Band Antennas

Multi-band antennas are intended to function effectively across several frequency bands. They are adaptable and can cover a wider variety of frequencies, enabling ham radio operators to converse on different bands without the need for additional antennas.

Types of Multi-Band Antenna

- **Dipole Antennas**: Multi-band dipole antennas are made up of many half-wave components for each band. They often feature traps or switches to isolate certain frequency bands.
- **Off-Center-Fed Dipole (OCF):** By feeding the antenna from a place other than the center, OCF dipoles enable multi-band capability. They might be tailored for particular frequency bands or a broad frequency range.
- **Yagi-Uda Antennas:** Yagi antennas can be constructed to cover many bands by using numerous driving elements and directors. The operator may pick the required band by switching between components.

Benefits of Multi-Band Antennas:

- **Space Saving**: When compared to having individual antennas for each band, multi-band antennas may save space.

- **Cost savings**: Since a single antenna can service several bands, extra equipment is not required.
- **Installation Ease:** Installing a single multi-band antenna is often easier than putting up many antennas.

Tunable Antennas

Tunable antennas enable the operator to alter the resonance or impedance of the antenna to fit the operational frequency. They often include variable components such as inductors or capacitors that may be changed for maximum performance at a certain frequency.

Tunable Antenna Types

- **Screwdriver Antennas:** These mobile antennas contain a motorized tuning mechanism that allows the antenna whip length to be adjusted. This enables simple tweaking while working on various frequencies.
- **Magnetic Loop Antennas:** Magnetic loop antennas use a wire loop with a variable capacitor to tune. They are portable and useful for work in confined spaces.
- **Tunable Vertical Antennas:** By varying the lengths of the radiating components, vertical antennas with adjustable elements may be adjusted to various bands.

Benefits of Tunable Antennas

- **Adaptability**: Tunable antennas can be fine-tuned for best performance on a given frequency, allowing for flexibility in changing operating circumstances.
- **Portability**: Some tunable antennas are intended for simple disassembly and travel, making them suited for mobile use.
- **Efficient Spectrum Use:** Operators may fine-tune the antenna to enhance efficiency on a specific frequency, decreasing interference and enhancing signal quality.

Antenna Matching and Tuning

What is an Antenna Tuner?

An antenna tuner, also known as a matching unit, is a radio communications equipment used to match the impedance of an antenna system to the impedance of the transmitter or receiver. It is a critical component in maximizing an antenna system's performance, guaranteeing optimum power transmission, and reducing signal loss. An antenna tuner improves signal reception and transmission by altering the impedance, resulting in greater overall communication quality. Antenna tuners are often used in amateur radio operations, where they play an important role in providing efficient and dependable communication.

Importance of Antenna Tuning

Antenna tuning is a critical component in optimizing an antenna system's performance. It is critical in maintaining maximum power transmission between the transmitter and the antenna, which results in increased signal intensity and overall system efficiency. Antenna tuners assist in decreasing signal reflections and standing waves by precisely matching the impedance of the antenna to the transmitter, decreasing interference and improving signal quality. Furthermore, antenna tuning enables accurate frequency selection, allowing the antenna system to work optimally across a broad variety of frequencies. As current communication systems get more complicated, mastering antenna tuners has become critical for ensuring dependable and high-quality wireless connection.

Types of Antenna Tuners

Various types of antenna tuners are routinely used. These tuners are essential in enhancing the performance of antennas by matching the impedance between the antenna and the transmitter or receiver. Manual, automated, and remote antenna tuners are the three categories. Human tuners need human tuning parameter change, but automated tuners can modify tuning settings automatically. Remote tuners, on the other hand, may be operated from a distance, making it possible to tune from a distance. Each kind of tuner has benefits and drawbacks, and the choice of tuner is determined by the antenna system's unique needs. It is essential to choose the correct kind of antenna tuner to guarantee an effective and dependable connection.

Understanding Impedance

In antenna tuners, impedance is a basic topic. It refers to the resistance of an electrical circuit to the passage of alternating current (AC). Understanding impedance is critical for transferring electricity efficiently between the transmission line and the antenna. Maximum power can be given to the antenna by matching its impedance to that of the transmission line, resulting in the best performance. The antenna design, physical size, and operating frequency are all important elements in determining impedance. Understanding impedance is critical for anybody wanting to improve the performance of antenna tuners.

Impedance Matching

An antenna tuner's impedance matching is critical. It entails modifying the antenna's impedance to match the impedance of the transmission line or the source. Maximum power transmission may be accomplished by impedance matching, resulting in better signal strength and decreased signal loss. Impedance matching requires careful consideration of variables like frequency, antenna design, and transmission line parameters. It is critical for maximizing an antenna system's performance and guaranteeing efficient signal transmission.

Impedance Mismatch

When the impedance of a transmission line or device does not match the impedance of the source or load, this is referred to as impedance mismatch. This may lead to signal reflections, power loss, and decreased system efficiency. Impedance mismatch in antenna tuners must be addressed to achieve optimum performance and maximum power transmission. Antenna tuners may decrease signal reflections and enhance power transfer by matching the impedance of the antenna to the impedance of the transmission line, resulting in better signal quality and overall system performance.

Antenna Tuner Basic Components

Variable Capacitors

Variable capacitors, which allow for accurate capacitance control, are a crucial component of antenna tuners. Antenna tuners may match the impedance of the antenna to that of the transmission line by adjusting the capacitance, guaranteeing maximum power transfer. These capacitors are built with a variable capacitance value that may be changed by spinning the rotor plates. Variable capacitor capacitance ranges generally vary from a few picofarads to several hundred picofarads. This large range of capacitance values enables the antenna system to be fine-tuned, enhancing its performance over a wide frequency range. Variable capacitors are widely employed in amateur and professional radio applications where precise impedance matching is essential for effective signal transmission.

Variable Inductors

In antenna tuners, variable inductors play a key role. They are critical in regulating the antenna system's impedance to ensure optimal power transmission. These devices enable for exact tuning of the antenna by adjusting the inductance, guaranteeing optimum performance over a wide range of frequencies. Variable inductors are very flexible and adaptable to varied antenna layouts due to their ability to fine-tune the inductance value. Whether the antenna is a dipole, a loop, or a Yagi, a well-designed variable inductance may substantially increase the tuner's matching capabilities, resulting in enhanced signal reception and transmission.

Balun Transformers

Balun transformers are an important part of antenna systems. They are essential in impedance matching, which ensures that the antenna and transmission line have the same impedance. Balun transformers aid in optimizing power flow from the transmitter to the antenna by matching the impedance, resulting in better signal strength and efficiency. Balun transformers are widely utilized in a variety of applications such as amateur radio, broadcasting, and wireless communication. They are classified into two types: voltage baluns and current baluns, with each built to meet certain impedance-matching criteria. Understanding balun transformer principles and applications is critical for mastering antenna tuners.

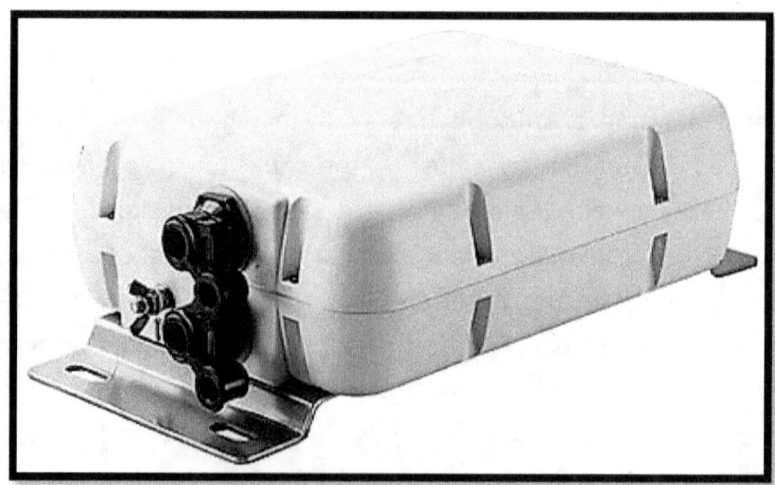

Tuning Techniques

Manual Tuning

Manual tuning is an important component of learning how to use antenna tuners. It enables precise modifications to be performed to achieve excellent antenna-to-transmission-line matching. Users may fine-tune the impedance matching by manually tweaking the antenna tuner to reduce signal loss and increase signal strength. To get the required resonance, different factors such as inductance and capacitance levels must be adjusted. Antenna tuners provide a great degree of control and flexibility with manual tuning, enabling customers to adjust their antenna system for the best performance.

Automatic Tuning

Automatic tuning is a critical function. It enables simple and exact antenna impedance matching to the transmitter or receiver. Automatic tuning removes the need for human adjustments by using clever algorithms and sophisticated circuitry, saving time and maintaining maximum performance. Antenna tuners can swiftly react to changing external conditions, such as temperature and humidity, with adaptive tuning, guaranteeing constant signal strength and signal loss reduction. This functionality is particularly useful for mobile and portable applications, where antenna impedance varies greatly. Overall, automated tuning improves antenna tuners' efficiency and efficacy, making them a crucial tool for obtaining optimum antenna performance.

Tuning with SWR Meters

Tuning using SWR meters is an important step in maximizing antenna tuner performance. Standing wave ratio meters, often known as SWR meters, are used to measure the impedance mismatch between the antenna and the transmission line. The impedance may be adjusted to reduce signal loss and increase power transmission by modifying the antenna tuner depending on the SWR values. Antenna tuners may employ SWR meters to fine-tune the antenna system to obtain the required resonance and increase overall signal quality. Understanding how to efficiently use SWR meters is critical for learning the art of antenna tuning.

Common Antenna Tuner Configurations

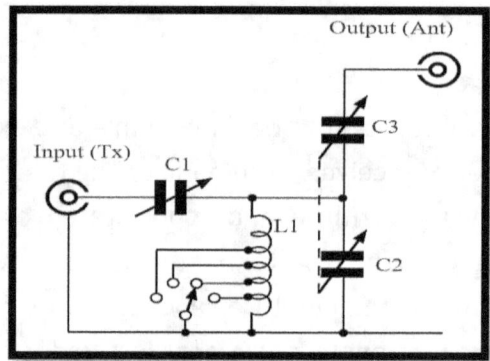

Pi-Network Tuner

A Pi-Network Tuner is a kind of antenna tuner found in radio frequency (RF) systems. It is intended to match the antenna's impedance to the impedance of the transmitter or receiver. A series capacitor, a shunt inductor, and a shunt capacitor make up the Pi-Network Tuner. These components are organized in a certain way that enables the antenna impedance to be adjusted. The Pi-Network Tuner is well-known for offering a broad variety of impedance-matching choices, making it a flexible solution for RF systems. It is often used in amateur radio, broadcasting, and telecommunications.

T-Network Tuner

T-network tuners are antenna tuners that are often used in radio frequency (RF) systems. It is intended to match the antenna's impedance to the transmission line's impedance, guaranteeing optimum power transfer and reducing signal loss. A series inductor and two shunt capacitors create a T-shaped circuit in the T-Network Tuner. This setup supports a broad range of impedance-matching capabilities, making it adaptable and appropriate for a variety of antenna designs. The T-Network Tuner is well-known among radio amateurs and professionals for its ease of use, efficiency, and efficacy in matching antennas.

L-Network Tuner

L-Network tuners are antenna tuners that are extensively used in amateur radio. It is intended to match the antenna's impedance to the impedance of the transmitter or receiver. This tuner is made up of two reactive components, an inductor and a capacitor, which are coupled in a certain way. The L-Network tuner is well-known for its ease of use and ability to match a broad variety of antenna impedances. It is often utilized in circumstances when other kinds of tuners cannot simply match the antenna impedance. The L-Network tuner is a vital tool for amateur radio operators wishing to maximize their antenna performance due to its ability to give a high degree of impedance matching.

Common Mode Current Baluns and Chokes

Differential mode

The differential mode is the typical mode of a transmission line (both coax and twin line) in which the currents in its conductors flow in opposing directions. A transmission line acts like two conductors in differential mode. On the interior of a coaxial cable, the currents are polarized, and there is no current flowing.

Common mode

A transmission line's common mode is when all of its conductor currents flow in the same direction. A transmission line functions as a single conductor in common mode. On the interior of a coaxial cable, currents flow in the same direction, and there is also a current flowing on the outside. The common mode along the feed line is a surface wave with an exponentially decreasing field in the radial direction. Depending on the dielectric interfaces involved, the phase velocity of a surface wave may be slower than the speed of light, equal to the speed of light, or even faster! Without a doubt, the group (energy) velocity can never exceed the speed of light.

Reasons to avoid using the common mode

Common mode currents on the feed line should be avoided for the following reasons:

- The common mode current route generates a stub at the antenna feed point, thus detuning your antenna.
- Furthermore, common mode currents at the antenna feed point modify the radiation pattern of your antenna.
- Common mode coax sheath currents will interfere with your electronic equipment (RFI) in the shack and during transmission.
- Common mode coax sheath currents contribute to the reception noise floor at the shack and during the reception, lowering the signal-to-noise level of desired signals.

Origin of the common mode

The contribution of two components results in the common mode:

1. The conducted common mode component extends the standing wave that runs the length of most antennas. (Aperiodic resistive antennas and leaky wave antennas are significant exceptions.) This standing wave is caused by the sudden change in conductivity and/or permittivity r at the antenna ends. In technical terms, such a sharp transition is referred to as a "**boundary condition.**"

2. The induced common mode component is caused by the feed line always being in the antennas near field. This means that enormous currents may be produced on the coaxial feed line's outer sheath, for example. Opposing induction fields cancel out and the generated common mode current is zero for symmetrical antennas fed in the middle. This is true for center-fed dipoles but not for off-center-fed dipoles (OCFDs), as NEC2 modeling demonstrates.

Antenna Input Impedances

The input impedances of antennas change between differential and common mode operation, as well as across antenna types. An antenna's differential and common mode input impedances can be calculated by treating the antenna as a two-wire, or single-wire, transmission line.

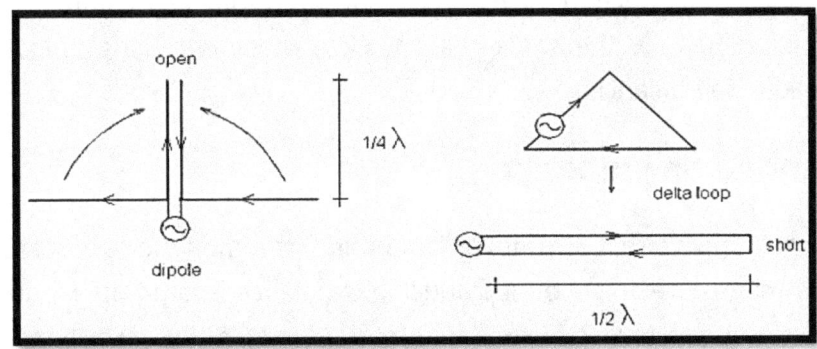

Differential mode antenna input impedance

When we speak about an antenna's input impedance, we usually mean its differential input impedance. It is the antenna input impedance calculated by considering the antenna as a lossy two-wire transmission line operating in differential mode. The loss is caused by radiation resistance. The legs of a half wave center fed dipole may be seen as a quarter wave long open wire transmission line in differential mode. At the feed point, this transmission line is open and translates this open across its lossy quarter wavelength to low differential mode input impedance. Similarly, a full wave loop antenna functions in differential mode as a lossy half-wave long open wire transmission line with a short at one end. At the feed point, this translates to low differential mode input impedance. A folded half-wave dipole acts similarly to a full wave loop, except that the radiation resistance is quadrupled owing to this antenna's added transformational effect.

Common mode antenna input impedance

An antenna's (conducted) common mode input impedance is calculated by considering the antenna as a single-wire transmission line in common mode. The legs of a half wave center fed dipole may be seen as a quarter wave long single wire transmission line in common mode. This transmission line is open at the end and changes this to low common mode input impedance at the feed point across its quarter wavelength. A complete wave loop antenna, on the other hand, functions in common mode like a half-wave long single wire transmission line with an open at one end. At the feed point, this results in high common mode input impedance. This explains why full wave antennas are less vulnerable to common mode noise reception. A folded half-wave dipole's common mode input impedance is also high for the same reason. Even more intriguing, folded half-wave dipoles are readily self-balancing.

Common mode on ladder line

Even via balanced open wire, twin line, or ladder line, common mode currents may flow. Naturally, this may be rectified by introducing a balanced common mode choke - or should it be termed a balbal, after the balun and unun? On the other hand, ladder line common mode currents may be handled in ways that coaxial cables cannot. For typical code currents, the balun or balanced tuner supplying the ladder line has a very high impedance. This high common mode impedance is changed to extremely low common mode impedance if the feed line is an odd multiple of a quarter wavelengths long. As a result, if the common mode antenna input impedance is also low, common mode currents will flow through the feed line. As a result, ladder lines should be kept at multiples of a half wavelength for antennas with a low common mode input impedance (e.g., a half-wave center-fed dipole). As a result, the antenna feed point will have a very high common mode impedance, effectively suppressing common mode currents on the feed line.

SWR Reduction Techniques

How to Lower SWR and Tune Ham and CB Antennas

Items Required to Set SWR

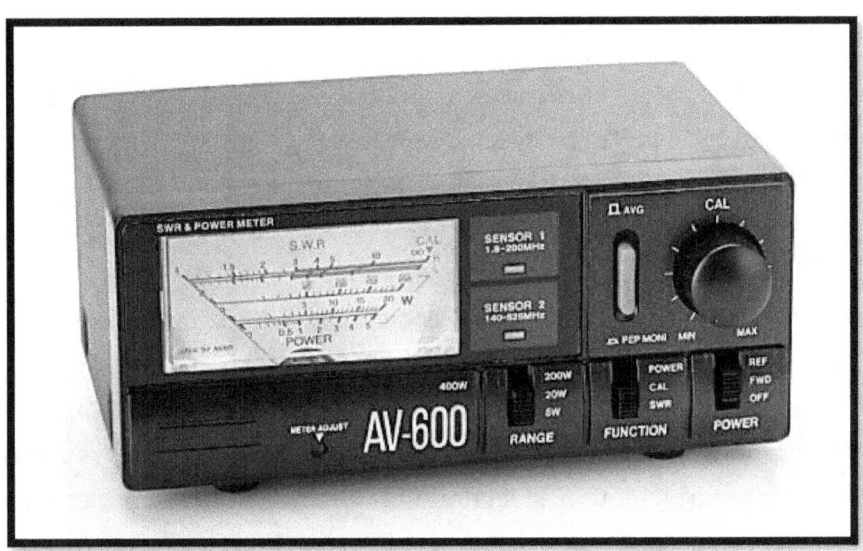

To adjust the SWR of your ham, GMRS, or CB antenna and transceiver, you'll need the following components.

Keep in mind that not all antennas need to be adjusted since some are already correctly tuned by the manufacturer.

- **SWR meter**: You will need an accurate SWR meter, such as the Surecom SW-33 MK2 or the Nissei RS-40, which we've used successfully. Depending on your transceiver type, you may need to acquire an antenna connection adapter and a short length of wire to connect to the output of your radio. Female PI-259 connectors are found on the rear of most ham radios and CBs.
- **Allen wrench set:** Many amateur radio and CB mobile antennas may be adjusted by releasing the little Allen screw that keeps them in place.
- **Bolt cutter**: A robust bolt cutter is required to cut a stainless steel antenna. (When using one, use eye protection.) You can also use a grinder to remove tiny pieces of antenna.

The steps:

Step 1: Determine the SWR of your radio and antenna

The first step is to examine the SWR of your radio and antenna to determine if any adjustments are required.

- Begin by parking your vehicle in an open place away from metal items.
- Connect your SWR meter between the transceiver and the antenna.
- Set the transmit power setting on your radio to "**low.**"
- Set the transceiver to the center of the amateur band you want to use, or to the CB or GMRS channel you intend to use.
- Check the frequency to verify if it is clear, and then provide your call sign (if appropriate) and declare that you will be testing, such as "**W4ABC**" testing.
- Before getting a reading, adjust your SWR meter if necessary. Otherwise, activate the microphone and record the reading. If it is less than 1.5:1, you may use your transceiver securely at that level and no more adjustments are required. (Lower is generally preferable, but be cautious when deleting any length of antenna, since this cannot be reversed!)

Step 2: Tune the Antenna

If you have a high SWR value, the next step is to adjust the antenna to be resonant at the frequency range you'll be using the most. You may simply need to shift the antenna up or down slightly without cutting it. However, in certain circumstances, you may need to cut some length off the antenna using your bolt cutters.

- Make a tiny line with a black permanent marker or pencil where the stainless steel antenna goes into the base.
- Loosen the Allen screw and lift the antenna element as much as possible, then tighten the screw.
- Check the SWR once again. If it is higher than it was before, the antenna element must be shorter to be resonant. If the reading is lower and you've pushed it up as high as you can, you're out of options since you can't add length to the element. A reading of less than 2.0:1 is considered safe for most transceivers, while lower is preferable.

- If the reading was higher when you lifted the element, try lowering it 1/16" below the black line and checking the SWR reading again. If it's getting lower, you're on the correct track. Write down the reading and continue lowering the element and repeating this step as needed.
- If your reading is 1.5:1 or below now, there is no need to continue; although a lower reading is usually preferable, if you begin chopping length off your antenna and go too far, you will have wrecked it. If your SWR is still more than 1.5:1, go to the following step.

Step 3: Reduce the Length of Your Antenna

If lowering your antenna reduces SWR and you've gone as far as you can but are still at 1.5:1, the next step is to shorten the antenna element.

- Place the antenna element in a vice and tighten it up.
- Using your bolt cutters (and using eye protection), gently remove no more than 1/2 cm or 13/6 of its length.
- Replace the element in the antenna base, drop it down, and tighten the Allen screw. Check the SWR again; if it's less than 1.5:1, you're done; otherwise, continue this procedure until you obtain the desired result.

Note that after you've reduced the length from your antenna, it can't be reversed, so continue with caution.

Other Factors That Can Contribute to High SWR

Aside from an antenna being excessively long or short, the following factors might generate high SWR readings.

- A short in the antenna cable or connector (a multi-meter may be used to test for this by looking for any shorting between the outside coaxial cable and the inside conductor.) There should be no conductivity between the interior and outside of your antenna cable.
- Antenna placement too near to the vehicle's cab or other metal items may occasionally result in a high SWR rating; moving the antenna away from the cab or other objects may resolve this issue.

- A buildup of bugs or road filth may be generating a short between antenna sections that are supposed to be isolated from each other; cleaning the antenna may resolve this problem.

SWR Meter Buying Guidelines

You must use an SWR meter designed for the frequency range of your ham radio, GMRS, or CB. CB radios run at 27 MHz, therefore you'll need an SWR meter designed for that frequency range. Since we have a GMRS radio as well as a dual-band VHF-UHF ham radio, we use the Surecom SW-102, which is designed to work from 125-525 MHZ. This is a good SWR meter for this range; however, it is not intended for use with CB radios. Another good "no tuning required" meter for VHF-UHF is the Nissei RS-40, which we've also used with success. It's best to choose an SWR meter that doesn't require calibration or tuning before each use, as this will ensure consistent readings as you perform the steps above in tuning your antenna. Models that don't require calibration also reduce the amount of unnecessary transmission over the airwaves during testing. Finally, make sure to get the connectors you'll need to connect the meter to your radio. The Surecom SW-102 has SMA connectors, so you'll need a pair of SMA female to UHF SO-239 adapters, as well as a short cable to connect the meter to your radio, to use it on most mobile transceivers.

Frequently Asked Questions

1. How do you lower SWR in Ham Radio?
2. What are the types of antennas and their features?
3. What are the different SWR reduction techniques?

CHAPTER EIGHT

BUILDING AND OPERATING A HAM SHACK

Overview

Want to learn how to build and operate a Ham Shack? This chapter will show you everything you need to know about building, designing, and operating a Ham Shack.

Designing your Ham Shack

To be able to take part in the hobby of amateur radio (also known as ham radio) and short-wave listening, the majority of individuals who are interested in these activities will want to establish their station. In the early days of radio, these rooms were referred to as radio shacks. This name has persisted and is still used today to refer to a room that houses the equipment that is used for ham radio. The majority of people may have quite varied requirements for their ham radio shack or amateur radio station. There is a wide range of possibilities available to them. The only thing that it may consist of for some people is a radio receiver that can be readily positioned in a corner of the room that is suitable for them. For those individuals who have a greater quantity of equipment, there will be a want for more room, and there may also be a requirement for wall maps and other charts.

Although this is the optimal option for many people, the ham radio shack does not necessarily have to be a whole room. Setting aside some space for the radio equipment may be accomplished in a variety of different ways. By using a little bit of inventiveness, it is possible to transform portions of the home that were not being used anymore into pretty nice shacks. To do this, it is required to first examine some of the fundamental criteria and then determine which sectors may be considered for conversion. Many other sites might be considered for the ham radio station. Some of these areas include spare rooms, loft spaces or attics, big and tiny cabinets, spaces in the garage, garden sheds, and a whole host of other spots. Each one has its own set of benefits and drawbacks, and with a little bit of forethought, it is often feasible to transform each one into a suitable location for the radio equipment.

Requirements for ham radio shacks

When it comes to the establishment of ham radio shacks or amateur radio stations, it will be required to reserve a certain amount of space for the equipment. The word "**radio shack**" has been used to refer to a ship's radio room or an amateur radio station ever since the early days of radio. This term has been used repeatedly since its inception. Many different pieces of equipment may be found in a standard amateur radio shack. These pieces of equipment include transmitters, receivers, and a range of other tools. Certain shacks are compact and tidy, holding just a limited amount of equipment so that they may be stored in a cabinet or the corner of a room. These shacks may be modest by design. Other individuals could be interested in reserving a space for their ham radio equipment. It is feasible to transform many different places of the house into an amazing ham radio shack with only a little bit of creativity and effort from the homeowner. Even garden sheds have been used well in this way.

It is important to keep in mind the following information, regardless of the location of the amateur radio shack:

- There must be sufficient provisions for mains electricity.
- The antenna feeders need to have access that is adequately suited to them. On the other hand, it should not grow too hot in the summer or too chilly in the winter.

- There should be adequate security measures in place, particularly in the case that the amateur radio shack is situated outside the main home, to prevent any equipment from being stolen.
- It is essential that the station be simple and convenient to run in its current position; for example, certain stations that are situated in a confined space inside a room may be difficult to use.
- It will be required to provide enough provisions for the construction of equipment

It is feasible to choose a site that is suitable for the amateur radio shack and to prepare for it to be simple to operate the equipment from there if one gives careful consideration to these and any other pertinent considerations. This is one manner in which the hobby of amateur radio may bring the greatest amount of pleasure and satisfaction.

Electrical wiring for the ham radio shack

Careful consideration should be given to the layout of both the main wiring and the lighting while the ham radio shack is being constructed and is being planned. When it comes to the wiring of the mains, it is essential to install a sufficient number of connections to feed the equipment that is now being used, while also allowing for some growth in the future. An appropriate solution is provided by the multi-way mains connection blocks that are sold in do-it-yourself shops. These blocks may be put below the back of the table. Cables may be properly routed out of the way so that they are not in the way. Nevertheless, it is essential to keep in mind that there must be adequate space beneath the top of the table to facilitate the movement of cables together with their connections in both directions. If there is not enough room, the connector may need to be detached from the cable to send it down to the mains socket, and then it could need to be rejoined thereafter.

When it comes to the installation of electrical wiring for an amateur radio station, safety must be placed at the forefront of the agenda and never overlooked. When taking into consideration the possibility that there is a large quantity of apparatus in the shack, it is essential to make certain that the circuits are not overloaded. It is important to get familiar with the wiring rules of the nation in question. The installation of a circuit breaker is another option that should be considered. It is common practice to include cut-outs that trip if any earth current or imbalance in line and neutral lines is detected using the equipment. Go to the rules once again to determine what is proper. In addition to this, it may be useful to have a single switch that can be used to separate the station from the rest of the equipment and turn it off when the station is not being used.

Lighting for an amateur radio station

Lighting is another problem that should not be overlooked by any ham radio station. The tabletop must be well-lighted if the shack is to be used to its full potential. If a single light source is provided from the center of the room, the surface of the table will perpetually be in the shadow of the individual who is engaged in the use of the apparatus. If any kind of building is carried out, this will be a very problematic situation. The best solution would be to have an angle lamp that, in addition to the lighting in the main room, can also be used to illuminate the area specifically designated for work. There is also the possibility of positioning a small strip lamp beneath a shelf that is placed on top of the table surface. However, a shade will be necessary to ensure that the lamp does not shine directly into the eyes of the person using it. In addition, be sure to follow the directions provided by the manufacturer for the fitting. An angle light, which can be used to illuminate any necessary region, is an option that might be considered. This may prove to be especially helpful if any kind of construction is planned, as it will be essential to make certain that the area where work is being done is adequately illuminated.

Buying the equipment

To be able to have the appropriate equipment is one of the most important considerations to make when establishing a ham radio shack. It should come as no surprise that there is a balance between space, performance, and cost. However, with some careful planning and investigation, it is typically possible to establish a radio station that is very effective while adhering to the constraints of the monetary and physical resources that are available.

Many different types of equipment might be required for the ham radio shack, including the following:

- **High-frequency (HF) transceivers:** There is a vast selection of frequencies (HF) equipment available. High-frequency (HF) transceivers typically cover frequencies ranging from 1.8 MHz to 50 MHz, which allows them to cover the lower end of the wide-frequency (VHF) spectrum. The vast majority of radio amateurs choose to go the route of purchasing commercially manufactured equipment because it is available in a wide variety of configurations. It is a significant undertaking to construct a high-frequency radio transceiver that meets all of the specifications, and sets can be acquired for a price that is very reasonable when one considers the contents of the set.
- **VHF and UHF equipment**: There is a wide selection of handheld and mobile VHF and UHF transceivers available for purchase. Generally speaking, they are for FM because a significant portion of the operation is mobile and portable, and FM is an excellent choice for this.
- **QRP equipment:** QRP, also known as low-power operation, gives individuals the opportunity to construct their various pieces of equipment. Because of its low power consumption and its typically more fundamental nature, the equipment can frequently be constructed, and there is a growing community of QRP enthusiasts who take pleasure in constructing and operating their equipment. Morse code is frequently used in operations because it makes the construction of the equipment significantly easier. Additionally, there are a great number of kits available that can provide an easy way to construct the equipment.
- **Station ancillary equipment**: The radio shack is equipped with a wide variety of equipment that can be used. These include antenna switches, which are used to manage the use of multiple antennas, VSWR meters, which are used to check the operation of the antennas, and power supplies, which are used to power the various rigs that are utilized.

Equipment layout

It is essential to pay attention to the arrangement of the ham radio equipment on the table. When the ham radio station is going to be used for extended periods, such as when it is being used for competitions, the ergonomics of the layout are of utmost importance. By placing the primary transceiver or receiver in the middle of the table, you will achieve the best results. It is no longer difficult to operate the tuning control while resting one's arm on the table. Also, the operation is significantly simplified over extended periods. It is possible to position additional large pieces of equipment on either side. It would not be difficult to install a linear amplifier or a second receiver in this location. One can choose to position the microphone on the left side of the device, which will free up the right hand for activities such as writing and taking notes. When it comes to the Morse key, it is possible to place it on the right-hand side of the table if it is being used. However, if someone is left-handed, it is obvious that these positions can be switched around. Additionally, there should be sufficient space on the table for a log book and a notepad. The notepad is extremely helpful for taking notes while the other station is talking, as well as for copying down Morse code.

Safety

The issue of safety is one of the most important considerations in any ham radio station. It is impossible to describe all of the features that ought to be utilized in this situation; however, we will provide a broad overview of some of the points that could be taken into consideration. The wiring of the mains is one of the most important aspects of an amateur radio station that affects the safety of the station. It is important to take into consideration many factors, including the incorporation of Residual Current Circuit Breakers (RCCBs), in addition to applicable wiring regulations. It is imperative that all wiring be completed to the highest possible standards. It is important to keep in mind that other people, especially children, also have the potential to enter the amateur radio station or ham radio shack, and they may not be aware of the risks involved. If there is even a remote possibility that children may enter the hut, it is preferable to make it as child-proof as possible.

Taking further measures includes ensuring that there are no potentially dangerous voltages in the vicinity. All soldering irons should be stored in holders at all times, and they should be turned off when there are other people around or when they are not being used. Having a broad awareness of safety is the first step that should be taken overall. Although it is very improbable that an accident would take place, the possibility of it happening may be minimized to the greatest extent possible by ensuring that all safety procedures are adhered to and that any potential dangers are removed as much as possible. The pastime may be enjoyed in a relaxed manner in this manner, with the knowledge that you and any guests who may enter the hut will not be in danger of any kind.

The Difference between Antenna Analyzers and Antenna Tuners

Different equipment that can be found in many ham stations is referred to by a variety of names, including antenna tuner, coupler, transmatch, matchbox, and ATU. Using this helpful equipment, amateur radio operators can assist in optimizing the connection between the transmitter and antenna, as well as improving the power flow between the two. Additionally, tuners can broaden the frequency range that antennas on a certain band can use. First things first: before we go any further, we must dispel a common misunderstanding: the antenna tuner does not tune your antenna or any component of it. Furthermore, it does not have any impact whatsoever on the SWR that exists between its output and the antenna. However, to make things easier, we will continue to refer to it as an antenna tuner since that is the term that is most often used in the ham world. To put it in the simplest words possible, an antenna tuner functions as a kind of impedance transformer that may be adjusted between the radio and the antenna. To match the output on the transceiver, it takes whatever impedance the antenna system provides it with and makes an effort to convert it to 50 ohms or anything that is relatively near to that value. Your transceiver can send its maximum amount of radio frequency power into the system when it detects an impedance of fifty ohms. In essence, the tuner gives the impression to your apparatus that it is broadcasting into a load with a resistance of fifty ohms, even if it is transmitting into a random wire or a pair of bedsprings.

On a resonant antenna, you do not need a tuner if you use a transmitter that operates on a single frequency or a cluster of frequencies that are located close to one another. To provide an example, the frequencies that encompass the 17-meter band span from 18.068 to 18.168 MHz, which is a rather modest portion of the spectrum. To cover the full band, a quarter-wave antenna that was cut for the center of the band would be sufficient. But take into account 80 meters. A staggering 4 MHz of bandwidth, ranging from 3.5 to 4.0 MHz, is covered by it. If you have a dipole resonant near the center of the band (3.8 MHz), you will discover that moving up or down the band will result in considerable alterations to the match between your transmitter and the antenna. For this particular scenario, an antenna tuner may be of assistance in using the antenna throughout a more extensive piece of the band. In case you make a few adjustments to the antenna tuner, your apparatus will nevertheless be able to detect an impedance of approximately 50 ohms. Tuners are also beneficial to antennas that do not need resonance. Through the use of an antenna tuner, it is possible to convert the vertical antennas measuring 31 and 43 feet, as well as the long wire and random wire antennas, into single or multiband applications. Although the tuner may be operated from your station, the optimal configuration for these antennas is to use a remote tuner that is attached directly to the antenna feed point. If you are in the market for a new transceiver, you should take into consideration the fact that many of the more recent rigs are equipped with built-in autotuners. Check the specifications of the transceiver or seek a button labeled "**TUNE**" that initiates the tuning process. The majority can withstand a minimum SWR of 3:1, which is enough for touch-up tuning and small mismatches. However, they are not capable of handling more severe mismatches, which is another reason why you also need to have a decent external antenna tuner.

Antenna Analyzer

An antenna analyzer is essentially a signal generator that is equipped with a variable frequency oscillator (VFO) that transmits a signal to the load that is coupled to it and a readout that shows a result set. In the process of constructing antennas, analyzers are often used by individuals. A physical tuning of an antenna and a check of the complete antenna system may be accomplished with the assistance of this instrument. You, the user, are the one who tunes an analyzer. With the assistance of the analyzer, you will be able to change the proportions of the antenna to get it as near to resonance as feasible. Analyzers often include visual displays, which make the data more straightforward to comprehend in comparison to a straightforward meter or digital readout. Some analyzers can generate many graphs for multiband antennas in a short amount of time and link to an external computer. If you have an antenna like a half-wave dipole, you can use an analyzer to modify it for the optimal SWR for a single range of frequencies, and you can also use it to find the available bandwidth that falls within the range of 2:1 or less. If you continue to operate on that frequency or within a particular bandwidth that is close to that frequency, the transmitter will detect impedance that is somewhat close to 50 ohms. On the other hand, the SWR values will increase if they leave that bandwidth. Because the majority of high-frequency antennas do not have a signal-to-noise ratio (SWR) that is flat over a complete band, many people choose a frequency that they like or tune for the middle of the band. Antenna analyzers are used by a significant number of amateur radio operators to resolve issues that are associated with antennas like the diagnosis of a break in coax cable or the identification of malfunctioning antenna components. It is possible that you could see unpredictable SWR readings at times, either because you will be unable to tune it correctly or because the tuning will not stay stable. These may be the result of a malfunctioning connection or the presence of water in the feedline. The feedline may be beginning to fail if your SWR measurements are consistently high over a while.

What is the Best Option for You?

In case you often construct or experiment with antennas, it is beneficial to have an analyzer on hand since it can be used to set up or troubleshoot antennas. In day-to-day operations, you will use a tuner to keep your transmitter happy by fooling it into believing that it is broadcasting into a load with a resistance of 50 ohms. In an ideal world with a liberal budget, it would be very beneficial to have both of these things. On the other hand, let's go over some instances of what each one performs and see whether they meet your requirements.

Custom Microphone and Headset Modifications

Custom microphone and headset modifications in ham radio may increase the audio quality, comfort, and usefulness of your communication equipment. Whether you want to increase audio quality, minimize background noise, or add more capabilities, there are various improvements you may explore. Keep in mind that changing your equipment may violate warranties, and it's vital to have a strong grasp of electronics and soldering abilities. Always observe safety standards and consider talking with experienced hams or electrical professionals if you're uncertain.

Microphone Modifications
- The element of the microphone should be upgraded so that it has improved sensitivity and frequency response. Improved audio clarity may be achieved by the use of electret condenser components of superior quality.
- If you want to minimize the amount of background noise that is present, put a noise-canceling circuit in your microphone. The incorporation of noise-canceling components or modules into the housing of the microphone is one method for doing this.

- To maintain a constant audio level, it is recommended that a dynamic range compression circuit be installed. This is very helpful for bringing the level of your voice into balance, especially in situations when there is a contrast between quiet and loud.
- To lessen the amount of low-frequency noise and handling noise, you need to include a low-cut filter. Eliminating undesired rumble and hum is one way that this might enhance the overall quality of the audio input.
- To improve the signal-to-noise ratio, the preamplifier of the microphone should be upgraded. This has the potential to result in greater performance as well as clearer audio signals.
- To get the best possible performance, adjust the impedance of the microphone so that it is compatible with the transceiver. Mismatched impedance may result in the loss of signal and a decrease in the quality of the audio.

Headset Modifications

- Replace the default ear cushions with choices that are of better quality and provide a greater level of comfort. It is possible to increase long-term comfort during longer operating sessions by using ear cushions that are filled with gel or memory foam.
- To improve the durability of the headset cable, consider replacing it with a braided cable. Adding connections that are compatible with your radio equipment is something you should think about doing to guarantee a safe and dependable connection.
- Increase the amount of noise isolation by using closed-back ear cups or adding more padding to the headphones. This may aid in boosting concentration during discussion by blocking out background noise and reducing distractions.
- Make the boom microphone movable so that you may arrange it in such a way that it provides the best possible voice pickup. This may prove to be very helpful in situations when you share the headset with other people.
- To conveniently alter the volume levels of the music without having to reach for the transceiver or an external audio device, you need to install an integrated volume control module.

- If you want to make your headset wireless, you should include a Bluetooth module. This will provide you with more flexibility of movement while its operation is taking place.

Frequently Asked Questions

1. How do you build your Ham Shack?
2. How do you design your Ham Shack?
3. How do you operate your Ham Shack?

CHAPTER NINE
SOFTWARE-DEFINED RADIOS (SDR) AND HAM RADIO

Overview

Chapter Nine emphasizes the need for software-defined radios and their integration with Ham Radio. Additionally, you will get to see SDR applications for Ham Radio.

Integrating SDRs with Ham Radio

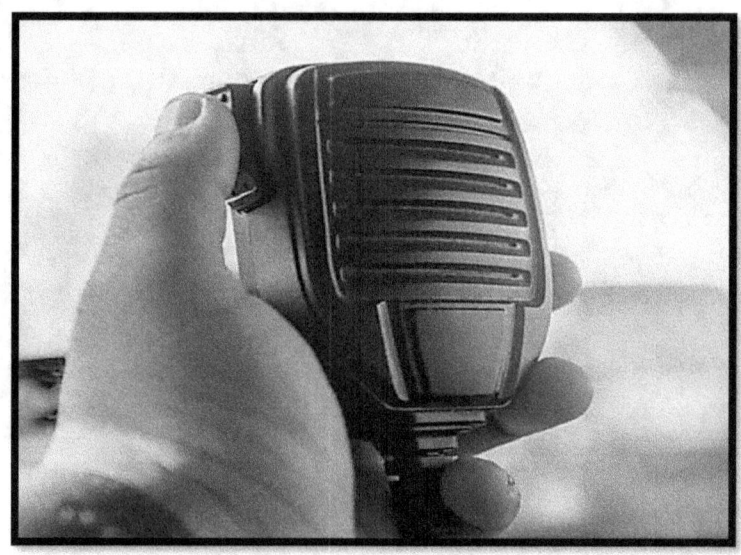

An increasing number of people are turning to software-defined radios. SDR-based radios are becoming more prevalent in both high-end and low-end applications as the cost of processing power continues to decrease, making it possible to integrate several software programs into use. One of the most significant advantages of super-resolution radio (SDR) technology is that it can be customized to precisely meet the needs of the user. With only a few simple software tweaks, the radio can be made to properly meet the requirements. Additionally, the integration of SDR with open-source programs like the GNU software is becoming less difficult.

What is a software-defined radio?

A radio that can be set to operate on a particular frequency is known as an SDR (Software Defined Radio). It is possible to alter the frequency of the radio (or the firmware) by the use of technology. When it comes to frequency bands or applications, hardware-based radios are often outfitted with specific components. Since the radio transceiver and modules at the front end are intended to operate at a certain frequency, it is not possible to readily modify the frequency of radios that are based on hardware. However, this is not the case with software-defined radios; these radios may be set to operate at a specific frequency of your choosing.

How does software-defined radio differ from software-controlled radio? What are the differences between the two?

An example of a type of radio known as software-controlled radio (SCR) is one in which software is used to control part or all of the processes that occur at the physical layer. To put it another way, this kind of radio is entirely dependent on software to monitor the radio's many fixed operations. Software-defined radio, on the other hand, is a kind of radio in which part or all of the functions at the physical layer are defined by software. This type of radio uses software to specify the functions. So to put it another way, the software is used to assess the capabilities and specs of the radio instrument. When the software on the radio is upgraded, there is a possibility that the radio's output and functionality will be altered.

The software that runs on SDR is based on a common hardware design. This architecture allows for the implementation of operations such as modulation and demodulation, filtering (which includes bandwidth changes), and other functions such as frequency selection and frequency hopping, if appropriate. Through the process of reconfiguring or altering the software, the output of the radio may be changed. To do this, software modules that can conduct radio operations such as broadcasting and receiving signals are used. These software modules are operated on a standardized hardware platform that comprises digital signal processing (DSP) processors and general-purpose processors.

What Benefits Does SDR Offer to Its Users?

- The chance to experiment with new things (for example, the liberty to create new protocols without interference).
- The elimination of analog hardware and the expenditures that are connected with it led to the development of radio designs that are less complicated and more effective.
- The capacity to detect and prevent infiltration from other communication networks.
- The capability to choose a frequency range and mode that is more appropriate for the circumstances that are now present.
- The capability to receive and transmit a variety of modulation schemes while making use of a common collection of hardware.

Significant Applications of Software-Defined Radios

The definition of the SDR software radio can be used in several contexts, including the following:

- It has been shown that radio amateurs have been able to make good use of software-defined radio technologies to enhance the device's efficiency and adaptability.
- The military has shown a significant amount of interest in software-defined radio technology, which enables them to recycle hardware and improve signal waveforms as required.
- Software-defined radio is beneficial to a great number of research initiatives. To give the precise transmitter and receiver requirements for each application, the radios may be configured to deliver those characteristics without the need to begin any process from scratch.
- There are several applications for software-defined radios, including mobile communications, which are particularly helpful. Without needing to update the hardware, it is feasible to make modifications to any specifications and even introduce new waveforms by upgrading the application. This requires no additional hardware to be upgraded. It is also possible to do this process remotely, which will result in considerable cost savings.

- SDR technology has the potential to be used in a wide range of additional applications, which enables the radio to be precisely tuned to the requirements via the implementation of software modifications.

Compatible SDR Hardware for Ham Applications

Through the use of software-defined radio hardware, short-wave radio (SDR) allows radio enthusiasts to experiment with various modulation schemes, frequencies, and signal-processing methods.

RTL-SDR Dongles

RTL-SDR dongles, which stand for Realtek Software Defined Radio, are not only inexpensive but also frequently utilized for SDR applications in the hobbyist community.

Frequency Range

Coverages often extend from 24 MHz up to 1.7 GHz in frequency.

Benefits include

- Economical in nature.
- Excellent for novices and those who want to experiment.

Disadvantages

- Limited bandwidth.
- A lower sensitivity in comparison to SDRs that are more specialized.

HackRF One

An SDR platform that is both more sophisticated and adaptable.

Frequency Range

It operates between 1 MHz and 6 GHz.

Benefits include:
- There is a broad frequency range.
- This product is suitable for a wide range of applications, including ham radio.

Drawbacks:

- A relatively greater cost.

Airspy R2:

A high-performance SDR that has exceptional sensitivity and dynamic range.

Frequency Range:

- It extends from 24 MHz up to 1.8 GHz.

Benefits include:

- In terms of sensitivity and dynamic range, the performance is exceptional.
- Capable of handling demanding applications properly.

Drawbacks:

The cost is higher in comparison to the cost of entry-level SDRs.

SDRplay RSPdx:

Featuring a high-performance receiving system design.

Frequency Range:

It extends from 1 kHz to 2 GHz.

Benefits include:

- Both the dynamic range and the sensitivity are exceptional.
- The presence of many antenna inputs allows for diverse reception.

Drawbacks:

The cost is higher in comparison to the cost of entry-level SDRs.

Advanced SDR Setup and Optimization

In ham radio, advanced SDR setup and optimization include configuring and fine-tuning the SDR gear and software for best performance.

1. Select an SDR that meets your needs. Think about things like frequency range, bandwidth, and dynamic range. High-end SDRs often provide greater performance, although financial constraints may impact your decision.
2. Using the included USB cable, connect the SDR to your computer. If necessary, connect the SDR to a steady power supply. Ground the SDR properly to limit the danger of interference.
3. Select SDR software that is compatible with your hardware. SDR# (SDRSharp), HDSDR, CubicSDR, and GNU Radio are all popular solutions. Follow the manufacturer's instructions for installing drivers.

4. Set the program to identify your SDR. Select the appropriate device and sample rate. Optimize the signal-to-noise ratio by adjusting the gain settings. Set gains are too high and you risk introducing too much noise.

5. Choose an antenna depending on your communication requirements (e.g., HF, VHF, and UHF). For optimal reception, place the antenna in an open location away from obstacles. Experiment with various antennas and positions to determine the best setup.

6. Use filters to restrict the bandwidth and reduce the possibility of interference. Match the bandwidth to the signal characteristics. A smaller bandwidth may increase the reception of weak signals.

7. Calibrate the SDR for precise frequency display. Some SDRs have calibration tools, while others may need manual adjustment.

8. Identify and minimize interference sources. Electronic equipment, power cables, and other radio transmissions are common causes. Experiment with various filtering and grounding approaches to reduce interference.

9. Investigate sophisticated signal processing methods including noise reduction and digital signal processing (DSP). To improve signal clarity, experiment with different demodulation options and DSP settings.

10. Connect your SDR setup to ham radio software such as WSJT-X for digital modes or FLdigi for multiple modes. Ensure that the SDR software and your preferred ham radio programs communicate seamlessly.

11. Look at remote operation possibilities if your SDR and software allow it. Securely configure remote access to control your SDR from many places.

12. Keep your SDR gear and software up to current with software and firmware upgrades. Participate in online forums and groups to share and learn from others' experiences.

13. Maintain thorough records of your configurations, settings, and successful installations. Documenting your experiences will aid in troubleshooting and replication of successful arrangements.

14. Make certain that your SDR operation adheres to local legislation and licensing requirements.

15. Learn about emergency communication methods using SDR. Prepare a backup power source and disaster measures.

Ham Radio SDR Applications

Waterfall Displays and Spectrum Analysis

Spectrum analysis and waterfall displays are critical ham radio tools, giving operators vital insights into the RF environment and facilitating effective communication.

Spectrum Analysis

Spectrum analysis entails visualizing and analyzing the frequency components present in a certain region of the RF spectrum. This is important in ham radio since it allows operators to determine available frequencies, interference, and signal intensities.

Equipment

- **Spectrum Analyzer:** A device that displays signal amplitudes at various frequencies. Portable and software-defined spectrum analyzers are extensively used in ham radio.
- **Computer**: Used to process and visualize data from a spectrum analyzer.

Applications

- **Frequency Selection**: Determine clear communication frequencies while avoiding interference from other communications.
- **Interference Detection**: Find and eliminate sources of interference that may interfere with communication.
- **Signal Monitoring:** For a better understanding of propagation circumstances, monitor signal intensities and variances over time.

Waterfall Displays

A waterfall display is a time-based graphical representation of spectrum analysis data. It shows how the RF spectrum evolves dynamically, with time on the horizontal axis and frequency on the vertical.

How Does It Work?

- **Color Mapping**: Different colors describe the amplitudes of signals, with brighter hues signifying stronger signals.

- **Time Axis**: Advances from left to right, displaying changes throughout time.
- **Frequency Axis:** Represents the frequency range under consideration.

Applications

- **Signal History**: Monitor changes in signal intensity and patterns over time.
- **Identifying broadcasts**: Identify new signals or broadcasts on the frequency band quickly.
- **Monitoring Band Activity**: Determine band occupancy and available frequencies.

The Importance of Ham Radio

- Spectrum analysis aids in the selection of appropriate communication frequencies, the reduction of interference, and the enhancement of signal quality.
- Waterfall displays assist operators in identifying occasional or intermittent interference, allowing them to take appropriate action.
- Tracking the spectrum over time enables operators to better understand propagation circumstances, allowing them to estimate the ideal periods for long-distance communication.
- Real-time data from spectrum analyzers and waterfall displays enables operators to immediately respond to changing RF circumstances.

Recording Functionality

1. SDR software often provides the option to capture sounds from certain frequencies. This may be used to record conversations, capture signals for study, or keep a journal of ham radio operations.
2. In addition to audio recording, some software allows for spectrum data logging. This contains signal strength, frequency, and modulation information, giving a full record of radioactivity.
3. For subsequent analysis and reference, recordings should contain timestamps and information such as frequency, mode, and signal intensity.
4. Recordings can be stored in a variety of file formats, including WAV and IQ (raw signal data). These files may be shared, archived, or processed using further software.

Applications

1. Real-time monitoring and recording help in the study of radio wave propagation circumstances, allowing operators to make educated judgments on the optimal communication frequencies.
2. Recorded signals can be used by hams to detect interference, weak transmissions, or strange propagation patterns.
3. The data collected can be used as a digital logbook, enabling operators to evaluate their radioactivity and conversations.
4. Real-time monitoring and recording are beneficial for teaching, allowing users to learn about radio communication, signal processing, and propagation.

Digital Modes and Signal Processing SDRs

SDRs (Software-Defined Radios) are becoming more popular in the ham radio community, allowing operators to experiment with a variety of digital modes and signal-processing methods. Digital modes in ham radio relate to the transfer of data in digital form, which allows for more efficient and resilient communication.

SDRs (Software-Defined Radios)

SDRs are radio communication systems in which software replaces or augments many of the usual hardware components, providing for more flexibility and adaptability.

Basic Components

- **ADC (Analog-to-Digital Converter):** Converts analog antenna signals into digital data.
- **DSP (Digital Signal Processor):** Demodulates, filters, and decodes digital data using software algorithms.
- **DAC (Digital-to-Analog Converter):** This device converts processed digital signals to analog signals for output.

Benefits in Ham Radio

- **Flexibility:** Through software upgrades, SDRs can handle a broad variety of frequencies and modulation methods.

- **Economical:** A single SDR device may replace numerous conventional radios for various bands and modes.
- **Upgradability:** Through software upgrades, new features and capabilities can be introduced.

Digital Ham Radio Modes

Types of Digital Modes

- **PSK (Phase Shift Keying):** Encodes data in the signal's phase.
- **RTTY (Radio Teletype):** For data transmission, audio frequency shift keying is used.
- **JT65/JT9:** Weak signal modes designed for low-power and congested band circumstances.
- **FT8/FT4:** Fast modes suited for fast exchanges in a short period.

The Role of SDR in Digital Modes

- SDRs can receive and transmit many digital modes using software-based modulation and demodulation operations.
- Software programs, which are often given by the SDR vendor or the ham radio community, make it easier to utilize certain digital modes.

Ham Radio Signal Processing

Demodulation and Filtering

- **Digital Filters:** SDRs use digital filters to remove undesired signals and noise, hence enhancing reception quality.
- **Algorithms for Demodulation**: Different algorithms are used to decode particular digital modes, turning the received signal into usable data.

Error Correction

- **Forward Error Correction (FEC):** Algorithms fix faults in received data, improving communication dependability under difficult settings.

DSP Techniques

- **Equalization**: Corrects signal distortions induced by transmission across various media.
- **Adaptive Noise Reduction**: Algorithms decrease background noise, resulting in a higher signal-to-noise ratio.

Software Tools and Platforms

SDR Software

- Popular SDR software includes SDR# (SDRSharp), GNU Radio, and HDSDR.
- These tools offer a graphical interface for customizing SDR parameters and viewing signals.

Digital Mode Software

- Applications like WSJT-X, fldigi, and MultiPSK make it easier to decode and encode numerous digital modes.
- Integration with SDR software enables easy mode switching.

Considerations and Challenges

- For typical ham radio operators, SDRs and digital modes may have a longer learning curve, needing knowledge of both hardware and software components.
- SDRs may encounter interference issues, particularly when employed in congested frequency bands. To prevent interfering with other services, ham operators must follow restrictions.
- Running advanced DSP algorithms and several software programs requires the use of a powerful computer, which may limit mobility.

Customizing SDR Software

By offering a flexible and adaptable communication platform, Software Defined Radio (SDR) has transformed the area of amateur radio (Ham Radio). Instead of depending on conventional hardware components, SDR software enables radio enthusiasts to interact with and alter signals directly via software. In Ham Radio, customizing SDR software entails modifying the functionality, look, and features to fit individual tastes and needs.

1. **Selecting the Best SDR Hardware**: Choose adequate SDR hardware before creating software. RTL-SDR, HackRF, and LimeSDR are all popular options. The capabilities and frequency range of your SDR system will be determined by the gear you use.
2. **Choosing SDR Software:** Select SDR software that works with your SDR gear. SDR software alternatives that are often used include:
- **SDR# (SDRSharp):** A popular and easy-to-use option for Windows.
- **GNU Radio**: A free and open-source platform for developing software-defined radios. It allows for more customization but has a higher learning curve.
- **HDSDR**: An additional Windows-based SDR program with a variety of functions.
- **CubicsDR:** An open-source, cross-platform SDR program with a contemporary UI.
3. **Installation and configuration**: Follow the installation instructions for the SDR software you've selected. Set up the program to detect and connect with your SDR gear.
4. **Changing the User Interface:** The user interface of many SDR software programs may be customized. Color palettes, window layouts, and text sizes may all be tweaked to improve visibility and usefulness.
5. **Filtering and signal processing:** Change signal processing parameters as bandwidth, filter type, and gain. Also, test several filters to increase signal quality and eliminate interference.
6. **Selection of Frequency and Band**: Create and store custom frequency presets for easy access to frequently used bands. Investigate the various frequency bands and modes available for your SDR device.
7. **Automation and custom scripting**: Advanced users should investigate scripting and automation possibilities. Scripting languages like Python are supported by several SDR systems. Create scripts to automate certain operations or to make real-time changes to signal processing.
8. **Data logging and recording:** Enable logging to record received signals, frequencies, and communication data. Set up data recording options to collect and analyze signals for later analysis.

Frequently Asked Questions

1. How do you customize SDR software?
2. How do you integrate SDR with Ham Radio?
3. What are the different compatible SDR hardware for Ham applications?

CHAPTER TEN
ADVANCED DIGITAL COMMUNICATION MODES

Overview

Chapter ten involves the need for advanced digital communication modes including high-speed data modes and weak signal communication.

High-Speed Data Modes

Exploring PSK-500 and Other High-Speed Modes

PSK is a digital modulation technology used extensively in ham radio communication. PSK-500 is a mode within the PSK family that is distinguished by its high data rate of 500 baud (symbols per second). In ham radio, enthusiasts often experiment with different modes to test different propagation circumstances, power levels, and antenna configurations. Let's look at PSK-500 and other high-speed ham radio modes:

PSK-500

Modulation Technique

- **Phase Shift Keying (PSK):** To represent digital information, PSK alters the phase of the carrier signal. When compared to lesser baud rate PSK modes, the phase of the carrier is altered in PSK-500 to communicate a larger data rate.

Advantages

- **Quicker and Higher Data Rate:** PSK-500 has a quicker data rate, making it ideal for efficient communication when bandwidth is available.
- **Robustness**: PSK is well-known for its resistance to noise and interference, making it excellent for low-signal circumstances.

Equipment and Software

- **Transceiver**: A PSK-modulated transceiver is needed. Modern ham radio transceivers often have built-in compatibility for several digital modes.

- **Computer**: A computer having a digital sound card interface. PSK-500 can be used with software like Fldigi, WSJT-X, or Ham Radio Deluxe.

Operating Procedures

- **Frequency Selection**: PSK-500 is commonly utilized in digital mode frequencies that have been authorized. Band planning and frequency etiquette should be followed by amateur radio operators.
- **PSK Software Configuration**: Configure the PSK-500 software, altering settings such as center frequency, audio volumes, and mode selection.
- **QSOs (connections):** Use PSK-500 to make on-air connections with other operators. Call signs, signal reports, and any extra information should be exchanged.

High-Speed Mode Considerations

- Ensure that band plans and rules for the use of high-speed digital modes on certain frequencies are followed.
- High-speed digital modes need efficient antennas. End-fed wire antennas, for example, are built for digital modes and may improve performance.
- Recognize how propagation circumstances impact the selection of high-speed modes. Certain modes may function better under certain meteorological conditions.
- Adjust the power levels to fit the communication distance needs. Maintain excellent amateur radio procedures and be aware of interference.

Setting Up High-Speed Data Stations

Setting up high-speed data stations entails using modern digital modes and technology to transfer data over the airways. This is generally referred to as digital communication because it enables ham radio operators to exchange words, photos, and even files via a variety of protocols.

Required equipment

Transceiver: Select a transceiver that can operate in digital modes. For connecting to a computer, most current transceivers feature built-in USB or sound card connections.

Computer: To run the digital mode program, you'll need a computer. Both Windows and Linux operating systems are widely used. Make sure your computer has a stable power supply since power outages might be troublesome during data transfer.

Sound Card Interface: An external sound card interface may be required to connect your transceiver to the computer. This interface converts audio impulses into digital data.

Antenna: An effective data transmission requires a good antenna. Consider using an antenna that is suited for the frequency ranges you want to utilize.

Coaxial Cable: Connect your transceiver to the antenna using high-quality coaxial wire. Signal loss may be reduced by using proper cables and connections.

Power Source: Ensure that both the transceiver and the computer have a consistent power supply. Uninterruptible Power Supplies (UPS) may assist in avoiding data loss in the event of a power outage.

Digital Mode Software: Select software that can handle high-speed data modes. Fldigi, WSJT-X, and JS8Call are examples of popular software. These systems often support a variety of digital modes like PSK31, RTTY, FT8, and others.

Data Interconnect Cables: Make sure you have the cords you need to connect your computer to the transceiver. USB, serial, and audio cables are examples of such cables.

Steps for Installation

1. Using the proper interface cable, connect the transceiver to the computer. Modify the transceiver settings to activate USB/LSB mode and configure the audio input/output levels.

2. Download and install the digital mode software on your computer. Set the software to match the settings on your transceiver. Configure the software's audio input/output settings to match your sound card interface.

3. Set up and configure your antenna for the appropriate frequency bands. Check that the antenna is properly grounded for safety.

4. Begin by broadcasting and receiving on low-power levels to test your setup. Check your signal strength and make modifications as required using the software's built-in features.

5. Experiment with several digital modes to discover the one that best meets your requirements and the current radio circumstances. Optimize performance by adjusting power levels, antenna design, and other variables.

6. Get to know the precise operational methods for the digital mode you're using. Maintain proper ham radio etiquette and abide by the laws and regulations.
7. Be ready to handle typical difficulties including interference, signal strength issues, and software setup mistakes. Update your software regularly to take advantage of bug fixes and upgrades.

Advanced Error Correction Techniques

Error correction is critical for guaranteeing dependable and clear transmission, particularly in difficult settings. Several innovative strategies are used to reduce mistakes and improve the overall performance of ham radio systems.

FEC (Forward Error Correction)

FEC is a technology that adds redundant information to send data to enable the receiver to identify and rectify mistakes without requiring retransmission. Ham radio operators often use FEC codes like Reed-Solomon codes and Turbo codes. Reed-Solomon codes are very useful for correcting burst mistakes, which are ubiquitous in radio transmission.

ARQ (Automatic Repeat reQuest)

ARQ is a protocol that is used to request the retransmission of data that has been received with errors. Selective Repeat and Go-Back-N are two commonly used ARQ methods. Selective Repeat retransmits just the packets that include errors, while Go-Back-N retransmits from the point where the problem occurred.

Interleaving

Interleaving entails altering the sequence of sent data to reduce the effect of burst faults. Interleaving enables FEC or ARQ methods to rectify mistakes more efficiently by spreading error bursts across a broader period of data.

Adaptive Equalization (AEQ)

Equalization compensates for communication channel distortion, especially in the presence of multipath propagation. Ham radios can use adaptive equalization algorithms, which continuously adjust filter parameters based on changing channel characteristics.

FHSS (Frequency Hopping Spread Spectrum)

FHSS transmits at rapidly changing frequencies, making it more resistant to narrowband interference and fading. By spreading the transmission across multiple frequencies, the impact of interference on overall communication is reduced.

OFDM (Orthogonal Frequency Division Multiplexing)

OFDM partitions a communication channel into multiple narrowband subchannels, each with its carrier frequency. OFDM is resistant to frequency-selective fading and allows efficient error correction techniques to be applied to each subchannel independently.

Adaptive Modulation

Adaptive modulation modifies the modulation scheme based on channel conditions. Ham radios can use feedback from the receiver to change the modulation scheme dynamically, optimizing it for the current conditions.

Diversity Techniques

Using multiple antennas at the transmitter and/or receiver to reduce fading effects. Spreading the same data across multiple time slots or frequencies to improve reliability in the presence of fading or interference.

CRC (Cyclic Redundancy Check)

CRC is commonly used to detect errors by appending a short checksum to the transmitted data. CRC and FEC can be combined to improve error detection capabilities.

SDR (Software-Defined Radio) and DSP (Digital Signal Processing)

SDR enables the implementation of advanced signal processing algorithms, such as sophisticated error correction techniques. DSP techniques can be used to improve error correction capabilities through adaptive modulation and coding.

Weak Signal Communication

The ability of amateur radio operators to communicate over long distances using very low power levels and highly efficient antenna systems is referred to as weak signal communication in ham radio. This aspect of ham radio is particularly intriguing because it requires operators to make the most of limited resources while overcoming the inherent difficulties of weak signal propagation. Modes such as CW (Continuous Wave), SSB (Single Sideband), and digital modes such as JT65, FT8, and PSK31 are commonly associated with weak signal communication.

- Weak signal communication frequently involves the use of low power levels ranging from a few milliwatts (QRP, or low power) to 100 watts. QRP is popular among hams that enjoy the challenge of making long-distance contacts with low power.
- Efficient antenna systems are critical for communicating with weak signals. Hams use a variety of antenna types, including wire antennas, vertical antennas, Yagi antennas, and loops. The frequency band, available space, and desired communication direction all influence antenna selection.
- Depending on the atmospheric conditions, different frequency bands behave differently. The HF (High Frequency) bands, such as 20 meters (14 MHz) and 40 meters (7 MHz), are frequently used for weak signal work. These bands have excellent propagation characteristics, making them ideal for long-distance communication.
- Weak signal communication uses specific propagation modes such as sporadic-E, meteor scatter, tropospheric ducting, and ionospheric conditions. These modes can boost signal strength and allow for longer-distance communication.
- Morse code, transmitted via CW, is a common mode of communication for weak signals. The ease of use of CW allows for efficient use of bandwidth and increases the likelihood of successful communication under difficult conditions.
- Another popular mode for weak signal communication is a single sideband. SSB consumes less bandwidth than AM (Amplitude Modulation) and consumes less power. Depending on the frequency band, operators will frequently use a lower sideband (LSB) or upper sideband (USB).
- Digital modes such as JT65, FT8, and PSK31 are intended for communication with weak signals. These modes use sophisticated error correction techniques and efficient

signaling to allow for reliable communication even when signals are weak and conditions are difficult.

- Hams pay close attention to their station setup when working with weak signals. This includes reducing RF interference, using low-noise amplifiers, and ensuring proper grounding to reduce unwanted noise and improve signal-to-noise ratio.

- It is critical to match the polarization of the transmitting and receiving antennas for efficient signal transfer. Horizontal and vertical polarizations are common options, and operators can change the polarization of their antennas to improve communication.

- Weak signal communication necessitates patience and perseverance. Operators may spend considerable time calling CQ (all stations) or patiently tuning for weak signals. When successful communication is established under difficult conditions, rewards follow.

- Some weak signal enthusiasts use beacon networks, in which automated transmitter stations transmit signals on specific frequencies indefinitely. This assists operators in assessing propagation conditions and identifying communication openings.

JT65 and JT9 Modes for Weak Signal Conditions

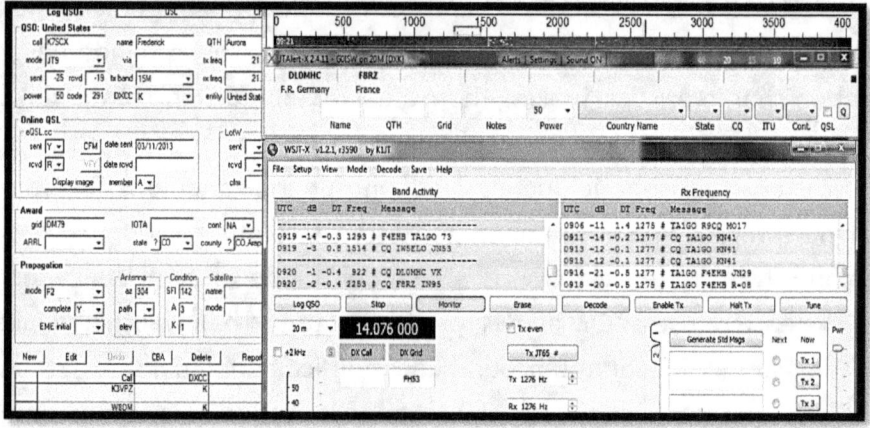

T65 and JT9 are digital modes particularly intended for amateur radio communication in poor signal environments. Joe Taylor, K1JT, created these modes, which are extensively used by ham radio operators to make connections in difficult propagation situations. Here's a more in-depth look at JT65 and JT9:

JT65

Goal and Design

- JT65 was developed for very weak signal transmission, particularly in ionospheric propagation circumstances that are often difficult for speech or Morse code.
- It is a slow, durable, and highly efficient digital mode that enables dependable communication even with weak signals.

Transmission Features

- JT65 broadcasts last 65 seconds and include 12 synchronization and data tones.
- Each tone is 47.6 Hz broad, and the signal's entire bandwidth is roughly 175 Hz.
- Due to the delayed transmission, signals considerably below the noise level may be deciphered.

Message Format

- Messages follow a specified format, which includes call signs, signal reports, and optional free-form content.
- The communication is often brief and to the point, making it ideal for fast information exchanges.

Error Correction

- Since JT65 uses extensive error-correcting methods, it is resistant to noise, interference, and fading.
- JT65 communication software can decipher signals even when they are hardly heard.

Usage

- JT65 is widely used in the HF bands, particularly at frequencies where signal transmission is poor.
- When other modes fail owing to poor propagation circumstances, operators often use JT65 to make long-distance communications.

JT9

JT65-like features

- JT9 and JT65 are quite similar in that both were created by Joe Taylor for poor signal transmission.
- JT9 is simply a faster variant of JT65, optimized for circumstances requiring a faster transmission of information.

Transmission Features

- JT9 broadcasts for 60 seconds and features nine 2-minute cycles within that time.
- Since each transmission cycle lasts 8.64 seconds, it is better suited for applications that need rapid exchanges.

Message Format

- JT9 messages, like JT65, have a defined format that includes call signs, signal reports, and optional free-form content.
- When compared to JT65, the shorter transmission time allows for a faster interchange of information.

Usage

- JT9 is often used on the same HF bands as JT65, although it is preferable for faster transmission.
- It's ideal for digital QSOs (conversations) in which the operators want to communicate information in a short period.

Decoding and software

- Both JT65 and JT9 modes need the use of specialist software for signal encoding and decoding.
- WSJT-X, JTDX, and other programs created by Joe Taylor and the amateur radio community are popular.
- The program provides automatic decoding, allowing operators to participate in weak signal communication more easily.

Techniques for Weak Signal Transmitting

Here are several weak signal-sending strategies often used by ham radio operators:

1. Use low-noise receivers to reduce the influence of background noise on weak transmissions. This contributes to a higher signal-to-noise ratio.
2. Select restricted bandwidth modes such as CW, JT65, FT8, and other digital modes. These modes are intended to work with extremely low signal levels and can decode signals that are barely above noise.
3. Use DSP technology to improve poor signals. DSP can be used to remove noise from incoming signals and increase their clarity. Many current transceivers have DSP capabilities.
4. Use high-gain antennas to boost both broadcast and received signal strength. Yagi antennas, for example, are very excellent at focusing a signal in a specific direction.
5. Use low-loss coaxial cables to reduce feedline losses. This aids in maintaining signal strength over extended distances.
6. Make certain that correct grounding is in place to prevent noise and interference. A solid ground may also help your antenna system work more efficiently.
7. Choose frequencies that are less congested and have less atmospheric and man-made noise. This is particularly critical when working with weak signals on the HF frequencies.
8. Choose to operate when there is less radio traffic. This decreases the possibility of interference and improves reception of weak signals.
9. FSK is a digital modulation technology that may be used to effectively transport data. FSK is used in modes such as RTTY (Radio Teletype), which might be useful for poor signal operations.
10. Boost weak signals at the receiving end using low-noise preamplifiers. However, take care not to introduce more noise into the system.
11. Monitor propagation circumstances and choose the best band for the time of day and current atmospheric conditions.
12. Use numerous receiving antennas to benefit from diversity reception. This can assist in reducing fading and improving overall signal reception.

Frequently Asked Questions

1. How do you set up high speed data stations?
2. How do you solve weak signal communication in Ham Radio?
3. What are the advanced error correction techniques in Ham Radio?

CHAPTER ELEVEN
HAM RADIO AND INTERNET INTEGRATION

Overview

Chapter eleven discusses about how to integrate Ham radio together with the internet to create a strong and successful connection.

Remote Operation via the Internet

Setting Up Remote Base Stations

Setting up distant base stations in ham radio entails many important procedures to enable reliable communication and regulatory compliance.

Understand Regulations

- Become acquainted with the local ham radio rules. Regulations may differ from nation to country, so make sure you are informed of and follow the unique regulations in your region.

Call Sign and License

- Obtain the required ham radio license for your area. To lawfully operate distant base stations, you must have a valid amateur radio license.
- During remote operations, use your allocated call sign.

Selection of Equipment

- Select the proper radio equipment for distant operations. Think about things like frequency bands, power output, and modulation types.
- Use a transceiver that can be operated remotely or a normal transceiver with a remote-control interface.

Remote Control Interface

- Purchase a remote-control interface that enables you to operate the transceiver from a distance. This may involve hardware solutions as well as software-defined radio (SDR) configurations.

- Popular remote-control interfaces include RemoteRig, FlexRadio Systems, and remote desktop software on a PC.

Antenna System

- Construct a strong antenna system for your distant base station. For best performance, consider antenna type, height, and direction.
- Make sure the antenna system is correctly grounded to prevent lightning strikes.

Remote Site Selection

- Select an appropriate location for your remote base station. Consider considerations such as accessibility, security, electricity, and internet connection.
- Ensure that any municipal zoning or land use restrictions are followed.

Power Supply

- Ensure that your remote base station has a consistent power supply. Depending on the locale, this may include a mix of mains power, battery backup, and perhaps other power sources such as solar or wind.

Internet Connection

- At a distant location, establish a robust and dependable internet connection. This is critical for controlling and monitoring your station remotely.
- For greater dependability, consider using redundant internet connections.

Remote Control Software

- To operate your transceiver via the internet, use remote control software. Software solutions vary, and certain transceivers may feature proprietary remote control software.
- Common software includes Ham Radio Deluxe, TeamViewer, and specialized solutions created by transceiver manufacturers.

Security Measures

- Put in place security measures to keep unauthorized people out of your remote base station. To protect the remote control interface, use strong passwords, encryption, and firewall settings.

Maintenance and testing

- Test your remote base station thoroughly before leaving it alone. This involves testing the radio, remote control devices, and internet access.
- Create a periodic maintenance program to guarantee the continued dependability of your remote setup.

Accessing and Controlling HAM Radio over the Internet

What happens when you mix the excitement of amateur radio with the power of the internet? The amateur radio community is worldwide, with members from every nation and culture. It's a business that has been nearly as long as electricity, with people using it to communicate all across the world long before the internet and mobile phones.

The internet has mostly dominated ordinary communication in the contemporary era. However, ham radios may be linked to wireless networks. Giving people from all around the globe the chance to listen in without the requirement for their ham radio equipment.

How to Listen to Ham Radio on Your Computer

There are hundreds of websites that allow you to listen to Ham radio online using your computer. To do this, they use two forms of technology: "SDR" and "RX," which can be accessed from your computer using the software listed below.

WebSDR

WebSDR is popular software that uses SDR technology to enable users to tune in independently. It is free and works well on Windows and Linux. WebSDR receivers can be found all around the globe, with many of them operated by local ham radio groups. To connect to WebSDR, just configure your chosen internet browser or SDR receiver program (software), such as the HRD Logbook.

It's a free, open-source program that lets you listen in on amateur radio discussions from anywhere in the globe. It is as easy as going to their website and putting in your location to use this service.

OpenWebRX

OpenWebRX, like WebSDR, is a free and open-source software package for Linux. It allows you to control SDR HF receivers remotely over the internet.

Is there a phone app that allows me to listen to Ham radio?

Yes, there are many Apps available; we suggest that you try the Echolink App. It's a free service that connects your smartphone to the Echolink system, allowing you to listen in real-time. It's available for both iPhone and Android phones.

Is a license required to listen to the ham radio online?

No, you don't need a permit to listen in. This is an excellent method to get started in the activity without investing any money. Meanwhile, it is crucial to understand that without a license, you cannot broadcast on ham radio.

How to Set Up Your Ham Radio for Computer Use

It is as easy to connect a ham radio to a computer as it is to attach an external USB speaker. When a ham radio is linked to a computer, it may be used for packet data, texting, phone calls, CW (Morse code), and even softmodem programming.

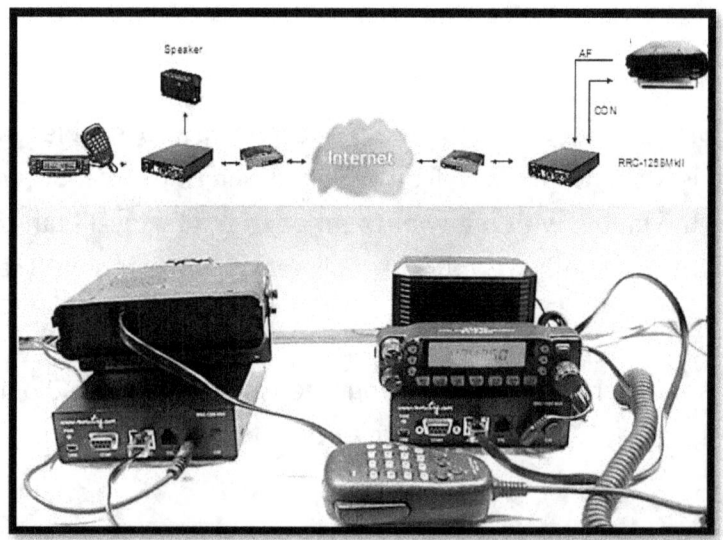

A computer, ham radio gear with a USB output (ideally), an antenna connector on the back of your PC, and necessary software to control/configure it is all required to connect your ham radio to a computer. A ham radio can be connected to the internet via a PC interface. There are various websites where you may listen to ham radio programs. These broadcast sites provide a range of quality levels, so if you have more bandwidth or want the highest possible sound quality, you may pick something with a higher quality level.

The following is a frequent modern-day solution:

- Run [PuTTY](putty.exe) to install a virtual serial port driver.
- Set your radio to data mode and connect an appropriate cable (typically included with the radio) to its microphone ("data") connector. This is accomplished by correctly setting the radio's "data" switch. You must be aware of the radio's specified mode (USB, LSB, etc.).
- In PuTTY, set the serial port settings to 19200 bps and "no flow control".

Connecting a ham radio to an Android smartphone necessitates the use of USB-to-something" or Bluetooth gear. Both ends will need relevant software.

Internet gateways and repeaters for amateur radio

A ham radio internet gateway is a system that routes speech, packet, or other data signals over the Internet. These are often linked back to local amateur repeaters in some form, allowing for far longer range than basic on-air FM conversations. It's worth noting that not all gateways support digital modes like packet and PSK31. To connect local ham radio repeaters via the internet, a gateway between your computer and the repeater is required; without it, GSM/Internet gateways are worthless. A GSM/Internet gateway is a device that allows a normal FM/D-Star transceiver to connect to distant Amateur Radio stations and repeaters using Internet Protocol networks such as the public Internet or commercial cellular data services.

The most popular sort of gateway is an IP radio, which enables you to connect your computer (by USB) or wireless router directly. To do this, just configure your PC's settings with the necessary information and allow Internet connection sharing on any additional network devices (routers/modems).

Analog repeaters are often used for FM communications, allowing you to communicate locally across a greater distance. To connect your ham radio to an analog repeater, just configure your PC settings and allow Internet connection sharing. There are several sorts of repeaters, the most common of which is an analog (FM) repeater. These are used for local FM communications and may be digitally coupled with GSM devices or software such as DStarRepeater.

Security Considerations for Remote Operation

Remotely operating a ham radio station necessitates additional security precautions to safeguard both your equipment and the broader integrity of the ham radio community.

Here are some important security issues for ham radio remote operation:

- Set up strict access rules for remote access to your ham radio station. To guarantee that only authorized personnel may manage the station remotely, use secure authentication mechanisms such as strong passwords or multi-factor authentication (MFA).
- Set up a strong firewall to keep unauthorized people out of your ham radio station. Only accept inbound connections that are required for remote operation. Update and patch your firewall regularly to address possible vulnerabilities. Secure connection between your distant location and the ham radio station by using virtual private networks (VPNs). This aids in the encryption of data and the prevention of eavesdropping.
- Encrypt the lines of communication between the remote operator and the ham radio station. This avoids unauthorized message interception and modification. For safe remote access, use protocols such as SSH.
- Update the firmware and software on your ham radio equipment regularly to fix vulnerabilities. Check that you are running the most recent software versions and that updates are gotten from approved and reputable sources.
- If you use remote control software to run your ham radio station, be sure it is safe and comes from a reliable source. This software should be updated regularly, and encryption should be used to safeguard communication between the control interface and the radio equipment.

- If feasible, safeguard your ham radio station's physical location to prevent unwanted entry. Locks, security cameras, and other physical security measures may be used.
- Create emergency shutdown protocols in the event of a security breach or other unforeseen event. Being able to quickly block remote access may help you avoid possible device damage or abuse.
- Set up logging and monitoring systems to keep track of remote access activity. Examine logs regularly for any unusual activity or illegal access attempts. This aids in the rapid identification and resolution of security concerns.
- If numerous people have access to remote operations, make sure they are all schooled on security best practices. Instill the significance of password security, identifying phishing efforts, and adhering to established security measures.
- Be aware of and follow any appropriate remote operating rules and guidelines in your location. This involves making certain that your remote operation adheres to radio frequency (RF) emission restrictions and other regulatory requirements.

Mesh Networking and HamWAN

Building Mesh Networks for Local Communication

Building mesh networks for local communication entails developing a decentralized network of radio nodes that can connect. Mesh networks are very beneficial when conventional communication infrastructure is inadequate or unavailable.

Understanding Mesh Networking

Familiarize yourself with mesh networking ideas. Each node in a mesh network serves as a relay, passing messages to other nodes. This decentralized structure improves network stability and expands network coverage.

Choose Ham Radio Equipment

Select ham radio gear that supports mesh networking protocols. Some radios and devices are purpose-built for mesh networking, but others may need firmware upgrades or extra hardware changes.

Choosing Mesh Networking Software

Select a mesh networking software or protocol that meets your requirements. AREDN (Amateur Radio Emergency Data Network) and HSMM-MESH are two common protocols. These protocols are intended for ham radio operators and may be loaded on devices that support them.

Node Placement

Plan where your mesh network nodes will go. To achieve enough coverage and connection, nodes should be strategically placed. Take into account terrain, obstructions, and the range of your equipment.

Node Configuration

Configure each mesh node using the program of your choice. Set the call sign, frequency, and other necessary characteristics for the node. Ascertain that each node in the mesh is set to communicate with surrounding nodes.

Antenna Considerations

To enhance the range and coverage of your mesh network, choose the proper antennas. Directional antennas may be used to connect nodes in a point-to-point fashion, but omnidirectional antennas can give larger coverage.

Power Considerations

Make certain that each node has a dependable power supply, particularly if you want to deploy nodes in distant regions. To ensure ongoing functioning, consider using solar power or battery backup solutions.

Testing and Optimization

Perform extensive testing on your mesh network. Identify and resolve any connection difficulties, and then optimize the location and configuration of nodes depending on the results of your tests.

Community Engagement

Engage with other ham radio operators if you're creating a mesh network for a particular ham radio community. Collaborate on node placement, exchange resources, and develop rules for mesh network usage and maintenance.

Training and documentation

Document your mesh network's setup and maintenance processes. Teach other ham radio operators in your area how to set up and troubleshoot mesh nodes.

Integrating HamWAN for High-Speed Data Links

Amateur Radio Emergency Data Network (HamWAN) is a high-speed data network created exclusively for amateur radio operators. It uses commercial off-the-shelf (COTS) technology and works in the microwave frequencies to offer ham radio hobbyists high-speed data connectivity.

Here are the key steps to integrate HamWAN into your Ham Radio setup for high-speed data links:

1. Learn about the exact frequency bands and hardware requirements for HamWAN. Using commercial Wi-Fi equipment, HamWAN commonly runs in the 5 GHz and 3.4 GHz frequencies.
2. Ensure that your usage of HamWAN conforms to your country's amateur radio rules. Check the exact frequency allotment for amateur radio operators as well as any supplementary high-speed data connection requirements.
3. Buy HamWAN-compatible gear, such as transceivers, antennas, and networking devices. HamWAN-compatible devices are often ordinary Wi-Fi equipment, however, it is critical to verify for HamWAN compatibility.
4. Configure the HamWAN transceivers and antennas at your operating site. For installation and setup, follow the manufacturer's instructions. For maximum performance, pay close attention to antenna alignment and polarization.
5. Connect to the HamWAN network by configuring your hardware with the proper network settings. This might include configuring IP addresses, gateway

information, and security settings. Make sure you have all of the appropriate credentials or permissions to connect to the HamWAN network.

6. Thoroughly test your HamWAN configuration to verify good operation. Examine the network's data transfer rates, latency, and general performance. Optimize the antenna alignment and other variables as required to attain the best outcomes.

7. Once your HamWAN configuration is complete, investigate the numerous applications that benefit from high-speed data connectivity. This might include data-intensive activities such as digital voice calls, video streaming, file sharing, and other data-intensive activities.

8. Join the HamWAN community to share your experiences, learn from others, and help the network grow. Participate in local HamWAN organizations or online forums to remain current on changes and best practices.

9. Since HamWAN was originally developed for emergency communications, consider how your system can be used in an emergency. Ascertain that you are aware of emergency procedures and that you can swiftly change your equipment to such instances.

10. Check for modifications to HamWAN specs, rules, and device firmware regularly. Maintain continued compliance with amateur radio laws and regulations by being updated about any changes that may impact your system.

Online Logging and QSL Management

Utilizing Online Logging Platforms

Online logging solutions can help amateur radio operators (ham radio operators) keep track of contacts, administer awards, and communicate their activities with the worldwide amateur radio community. Here are some examples of how to use online logging systems in ham radio:

LOTW Electronic Logging System (eQSL)

- **eQSL (Electronic QSL Card Centre):** eQSL is a website where hams may exchange electronic QSL cards. It offers an easy method to confirm and maintain contacts online.

- **LOTW (Logbook of the World):** Another electronic QSL system sponsored by the ARRL is LOTW. It enables operators to submit records, electronically validate contacts, and receive incentives.

QRZ.com

- **Logbook**: QRZ.com, a prominent ham radio website, has a logbook function that allows you to record your contacts online. It also has capabilities for examining log data.

Club Log

- **DXpedition and Contest Logging**: DXpedition and contest logging are popular uses for Club Log. It includes capabilities like online log searching, QSL matching, and others. It's particularly beneficial for keeping track of progress toward certain rewards.

Ham Radio Deluxe

- **Integrated Logging Software**: Ham Radio Deluxe is a software package that incorporates logging, rig control, digital modes, and other features. The logging component keeps track of operators' contacts and interfaces with numerous web platforms.

N1MM Logger +

- **Contest Logging**: Many hams use N1MM Logger+, a popular contest logging program. It supports a broad variety of competitions and includes features such as real-time scoring. While it is essentially a local application, logs may be exported to web platforms for validation and analysis.

Online ham radio communities

- **Forums and Social Media**: Sites like Reddit's r/amateurradio, different ham radio forums, and social media groups often feature areas for sharing logs, discussing contacts, and exchanging information.

Cloud-Based Logging Platforms

- **Hamlog.eu, HamQTH, and others:** There are internet logging systems such as Hamlog.eu and HamQTH. These systems provide cloud-based logging services, allowing you to view and edit your log from any location.

Efficient QSL Management and Confirmation

QSL cards are used to confirm two-way radio communication between amateur radio operators, and they must be managed properly to keep correct records and confirm interactions.

1. Use electronic QSL services like eQSL.cc and the ARRL's Logbook of the World (LoTW). These services enable hams to electronically exchange QSL confirmations, expediting the confirmation procedure.
2. Keep an online logbook with sites like QRZ.com, HamQTH.com, or ClubLog. These systems often interact with eQSL services and provide simple log synchronization.
3. Use logging software that allows for automated submission to electronic QSL systems. This may save time and lessen the likelihood of log entry mistakes.
4. Join a QSL bureau to exchange actual QSL cards with other operators all over the globe. Bureaus aid in the efficient management of card distribution.
5. When sending QSL cards by direct mail, include a self-addressed, stamped envelope (SASE) if you like to receive physical QSL cards. This encourages the other operator to send back a QSL card.
6. Some operators hire a QSL manager to distribute QSL cards on their behalf. This is particularly typical for DXpeditions or stations with considerable traffic.
7. To save time while producing QSL cards, print labels with your QSL information. This is very helpful for direct QSLing.
8. Define your confirmation rules clearly and communicate them to other operators. Specify if connections are automatically confirmed via electronic methods or whether QSL cards are only sent upon request.
9. Regularly backup your log data to prevent loss of information. This is essential for keeping correct records and addressing conflicts.
10. Join systems like GlobalQSL or similar services that assist in streamlining the QSL confirmation procedure, particularly for direct QSLing.

Frequently Asked Questions

1. How do you set up remote base stations?
2. How do you integrate Ham Radio and the internet?
3. How do you build mesh networks for local communication?
4. How do you use online logging platforms?
5. How do you access and control Ham Radio over the internet?

CHAPTER TWELVE
HAM RADIO TRANSVERTER AND MICROWAVE OPERATIONS

Overview

Chapter twelve talks about Ham Radio transverter and microphone operations including microwave bands and equipment and earth moon earth communication.

Transverter Operation and Integration

Enhancing Band Coverage with transverters

Transverters are devices used in ham radio to enhance the frequency coverage of a transceiver by transforming signals from one frequency to another. They are very beneficial for increasing band coverage, enabling you to operate on bands that your transceiver may not support natively.

Here are some steps and considerations for improving band coverage in ham radio using transverters:

Transverters are usually made up of a mixer, a local oscillator (LO), and filtering components. They transform signals from one frequency band (for example, microwave bands) to another (for example, HF or VHF bands).

Choosing the Best Transverter

- Select a transverter that covers the frequency range you wish to investigate.
- Think about things like input/output frequency, gain, and compatibility with your transceiver.

Frequency Planning

- Understand your transceiver's frequency range as well as the extra bands you wish to cover with the transverter.
- Consider your operating frequencies and bands carefully.

LO (Local Oscillator) Frequency

- Adjust the transverter's local oscillator frequency to match the conversion required for the target band.
- For optimal performance, ensure that the LO frequency is steady and has minimum phase noise.

Connection and Integration

- Using the appropriate wires, connect the transmitter to the transceiver.
- To integrate the transverter with your transceiver, follow the manufacturer's instructions.

Antenna Considerations

- Use an antenna appropriate for the transverter's frequency range.
- Make sure the antenna system is correctly matched and calibrated for peak performance.

Building and Calibrating Transverters

Building a Transverter

1. Determine the bands you want to work on and select acceptable frequencies. Understand your transceiver's specs to guarantee compatibility.
2. Choose a transverter design that meets your needs. There are several designs accessible online, as well as kits.
3. Purchase or acquire the required transverter components, such as mixers, filters, oscillators, amplifiers, and suitable housing.
4. Make or use a schematic for your selected design. Create a plan for your transverter, taking adequate grounding and RF shielding into account.
5. Put the circuit together on a printed circuit board (PCB) or a prototype board. Carefully solder components to ensure correct connections.
6. Test each level of the transverter to confirm that each component functions properly. Examine the signal for any unwanted signals or harmonics.
7. Connect all of the components and test the transverter as a whole.
8. Enclose the transverter properly, with adequate grounding and shielding.

Calibrating a Transverter

1. Check the transverter's output frequency using a frequency counter or a calibrated receiver. Tune the local oscillator until the frequency matches the desired output frequency.
2. With a power meter, determine the output power. Make necessary adjustments to the amplifiers and attenuators to reach the appropriate output power level.
3. Check the transverter's standing wave ratio using an SWR meter. Modify matching networks and filters to reduce SWR.
4. Align your transverter's filters appropriately to guarantee the best performance.
5. Consider the effects of temperature on the transverter components and make any temperature compensation modifications that are required.
6. Check that the transverter is sensitive enough to pick up on weak signals. To acquire the required sensitivity, adjust the IF gain and other settings.
7. Examine the circuit for intermodulation distortion and make any necessary adjustments to reduce it.

Challenges and Solutions in Transverter Use

Here are some typical issues and possible remedies when utilizing transverters in ham radio:

Frequency Drift

- **Difficulty**: Transverters might suffer from frequency drift over time, compromising signal stability.
- **Solution**: Use transverters with temperature-compensated oscillators that are stable. Allow the transverter to warm up before using it to reduce drift.

Frequency Calibration

- **Difficulty**: Accurate frequency calibration may be difficult, particularly when utilizing various transceivers and transverters.
- **Solution**: Use a recognized and reliable frequency reference to calibrate the transverter. Calibration options are integrated into certain transverters, and external frequency counters or GPS-disciplined oscillators may be utilized.

Interference

- **Difficulty**: Transverters might be affected by interference from neighboring electrical gadgets or radio transmissions.
- **Solution**: Use high-quality coaxial wires and ensure sufficient transverter shielding. Locate the transverter away from interference sources and consider employing filters to reduce undesired signals.

Power Supply Noise

- **Difficulty**: Power supply noise might introduce undesired signals and degrade transverter performance.
- **Solution**: For the transverter, use a clean and reliable power source. To decrease power supply noise, consider installing filters or regulators.

Compatibility Issues

- **Difficulty**: It might be difficult to ensure compatibility with transverters and transceivers from various manufacturers.
- **Solution**: Use known compatible transverters and transceivers, or interface devices that offer adequate impedance matching and signal level conversion.

Amplifier Integration

- **Difficulty**: Integrating transverters with power amplifiers may be difficult due to impedance matching and power levels.
- **Solution**: Use amplifiers built expressly for use with transverters. To reduce signal losses, ensure adequate impedance matching and utilize low-loss coaxial cables.

Weather and Environmental Considerations

- **Difficulty**: Outdoor transverter installations may encounter difficulties due to weather and environmental circumstances.
- **Solution**: For outdoor installations, use weatherproof enclosures and connections. To manage temperature changes, consider extra protective measures such as heat sinks or cooling fans.

Working in Split Mode

- **Difficulty**: When using transverters, operating split frequencies might be more complicated.

- **Solution**: Learn about the split operating characteristics of your transceiver and transverter combo. Check that the transceiver and transverter are both properly set for split operation.

Microwave Bands and Related Equipment

Exploring the Microwave Frequency World

Exploring the world of microwave frequencies as a ham radio operator may be both fascinating and hard. Ham radio operators, sometimes referred to as amateur radio operators, have access to a broad variety of frequencies, including those in the microwave spectrum. Consider the following crucial points:

Frequency Ranges

Microwave Bands: Microwave bands are defined as frequencies over one gigahertz (GHz) and can extend into the millimeter-wave region. 2.3 GHz, 3.4 GHz, 5.7 GHz, 10 GHz, 24 GHz, and higher are common microwave bands utilized by hams.

Equipment

1. **Transceivers**: Microwave bands need the use of specialized transceivers. To work successfully at higher frequencies, these transceivers often include frequency converters, mixers, and amplifiers.
2. **Antenna**: Due to the shorter wavelength, microwave antennas are tiny and extremely directed. Parabolic dish antennas are often used for signal concentrating.
3. **Feedhorn and Low-Noise Amplifier (LNA):** To amplify weak signals at microwave frequencies, a feedhorn (a device that collects the signal and delivers it to the antenna) and a Low-Noise Amplifier (LNA) are often used.

Regulations and Licensing

1. **Licensing**: Operating on microwave frequencies may need extra licenses or authorization. Make sure you are familiar with the laws in your nation or area.
2. **Power limitations**: Power limitations for various microwave bands may vary, and operators must comply with these limits to prevent interference and legal difficulties.

Propagation

1. Due to their higher frequencies, microwave transmissions are often confined to line-of-sight communication. Signal transmission may be hampered by obstacles such as buildings and topography.
2. **Weather Conditions**: Weather conditions, particularly rain, may weaken microwave transmissions. When planning and running, these elements must be taken into account.

Weak Signal Operation

1. **Weak Signal Modes:** For weak signal operations on microwave bands, Morse code (CW), digital modes like JT65 and FT8, and different digital signal processing (DSP) methods are often used.
2. **Beacons**: Many microwave enthusiasts put up beacons to send signals continually, allowing others to test and improve their equipment.

Safety

1. **Safety Precautions**: When using high-frequency equipment, always take care. Be cautious of high-power broadcasts and associated RF risks.
2. **Equipment Safety**: Check to see whether your equipment meets safety regulations, particularly if you're working with greater power levels.

Building and Modifying Microwave Equipment

Building and upgrading microwave equipment may be a difficult but rewarding task. Microwaves are commonly defined as frequencies over 1 GHz, with ham radio bands typically ranging from 1.2 GHz (23 cm) and upwards.

The steps include:

1. Be sure you grasp RF (Radio Frequency) fundamentals, microwave theory, and safety measures.
2. Obtain the required amateur radio license to operate in the microwave frequencies. Different countries have different standards for licensure.

3. Select the microwave band in which you wish to operate. 23 cm (1.2 GHz), 13 cm (2.4 GHz), 9 cm (3.4 GHz), 6 cm (5.7 GHz), and 3 cm are all common ham radio microwave bands.

4. Purchase or construct transceivers, antennas, and other essential equipment. Components for microwave frequencies may be more difficult to get than those for lower frequencies.

5. Create or acquire high-gain antennas for microwave frequencies. Microwave bands often use parabolic dish antennas.

6. Create or edit transceivers for the selected frequency band. This might include building or altering RF amplifiers, mixers, and oscillators.

7. Filters can be used to reduce undesirable harmonics and signals. Microwave filters are critical for transmitting clean signals.

8. As power needs grow with higher frequencies, develop or alter power amplifiers to provide the required output.

9. Discover the characteristics of waveguides and coaxial cables at microwave frequencies. To avoid signal loss, properly match and terminate your transmission cables.

10. Keep safety measures in mind, particularly while dealing with high-frequency equipment. Microwaves may be hazardous to one's health; hence care should be taken to prevent exposure.

11. To test and calibrate your equipment, use suitable test equipment such as spectrum analyzers and network analyzers. At microwave frequencies, alignment is crucial.

Microwave Band Operations: Techniques and Challenges

Operating on microwave bands entails using frequencies exceeding one gigahertz (GHz), often in the UHF (Ultra High Frequency) and SHF (Super High Frequency) ranges. Microwave bands can support high data rates, narrow beams, and complex modulation methods. They do, however, create distinct obstacles for amateur radio operators.

Here are some strategies and issues related to ham radio operation on microwave bands:

Techniques

Antenna Design

- **High-Gain Antennas:** High-gain antennas are used to concentrate a signal in a certain direction, accounting for the increased free-space path loss at microwave frequencies.
- **Parabolic Reflectors**: To obtain high directionality and gain, use parabolic reflector antennas such as dish antennas.

Frequency Coordination

- **Coordination with Local Authorities**: To minimize interference, coordinate your frequency use with local authorities and other amateur radio operators, particularly in heavily populated regions.

Propagation Considerations

- **Line of Sight (LOS):** Since microwave transmissions are predominantly line-of-sight, your station location and antenna heights should be planned properly.
- **Fresnel Zone Clearance**: Maintain a clean Fresnel zone (the region surrounding the visual line of sight) to reduce signal interference from objects.

Equipment Selection

- **High-Quality Transceivers**: As the requirement for signal purity increases at higher frequencies, invest in high-quality transceivers with steady and clean transmissions.
- **Low-Noise Amplifiers (LNAs):** Low-noise amplifiers are used to amplify weak signals while introducing little noise.

Digital Modes and Modulation

- **Digital Modulation**: For effective bandwidth usage and increased signal integrity, consider employing digital modulation algorithms such as frequency-shift keying (FSK) or phase-shift keying (PSK).
- **Error Correction**: Use error correction methods to improve communication dependability.

Weather Considerations

- **Rain Fade**: Rain fade is a phenomenon in which microwave transmissions are dampened by precipitation. This may have an impact on signal strength during inclement weather.

Challenges

- Since microwave signals have greater free-space path loss than lower frequency bands, they need more power and gain to communicate across longer distances.
- Microwave equipment is often more costly than lower-frequency alternatives. This figure covers the cost of transceivers, antennas, and specialist components.
- At higher frequencies, compliance with spectrum restrictions becomes more crucial. Make sure you're operating inside amateur radio bands and within power limitations.
- Due to the narrow beams associated with higher-frequency antennas, achieving and maintaining exact antenna alignment is critical for efficient transmission.
- Compared to the HF and VHF bands, microwave ham radio equipment has comparatively limited commercial support, necessitating operators to often construct or upgrade their gear.
- Interference from other sources, such as other electronic equipment and communication networks, may occur in microwave bands. Filtering and shielding are critical.

EME (Earth-Moon-Earth) Communication

Setting Up for EME Contacts

Moonbounce, or Electromagnetic Environment (EME), is an interesting feature of amateur radio that includes bouncing signals off the Moon to create long-distance conversations. Setting up for EME connections needs careful preparation, specific equipment, and a thorough awareness of the particular obstacles that this style of communication presents.

Here's a brief overview to get you started:

Equipment

Antenna:
- Use a directional antenna with a limited beamwidth and high gain.
- For EME, Yagi antennas with several elements are typically used.
- Circular polarization is often used to reduce signal fading.

Transceiver:
- A high-power transceiver, ideally with at least 100 watts output, is required.
- The most popular mode for EME is SSB (Single Sideband), although CW (Continuous Wave) is also used.

Amplifier:
- Consider using a power amplifier to amplify your signal, since high power is required for Moonbounce communication to be effective.

LNA (Low-Noise Amplifiers):
- Low-noise amplifiers (LNAs) are essential for receiving weak signals. Low-noise preamplifiers should be used at the antenna feed point.

Antenna Placement

Precise Alignment:
- It is critical to aim accurately towards the Moon. Calculate the Moon's location using tracking software.

Azimuth and Elevation:
- Since the Moon moves in both azimuth and elevation, a rotator that can do both is required.

Frequency Planning

Choosing a Frequency:
- Select a frequency in the amateur radio bands that are typically used for EME.
- Frequencies like 144 MHz, 432 MHz, and higher bands are often used.

Doppler Shift:
- Take into account the Doppler shift caused by the relative velocity of your station, the Moon, and the Earth. Many transceivers and tracking software are capable of doing this automatically.

Operating Procedures

Calling CQ:

- When calling CQ, move slowly and deliberately to allow others time to reply.

Information Exchange:

- A standard exchange usually consists of a signal report, a grid square, and a sequence number.

Patience and perseverance:

- Due to signal propagation delays, EME connections may take some time. Be persistent and patient.

Advanced EME Operations Techniques

Earth-Moon-Earth (EME) communication, commonly known as moonbounce, is a unique and difficult component of ham radio that includes using the moon as a natural satellite for signal reflection.

Here are some advanced approaches for ham radio EME operation:

1. Concentrate sent and received signals by using high-gain antennas with narrow beam widths. This aids in directing energy toward the moon and improving the signal-to-noise ratio.

2. Circular polarization aids in the prevention of signal depolarization during reflection. Circularly polarized antennas are less susceptible to Faraday rotation, which may increase signal dependability.

3. At the antenna feed point, use low-noise amplifiers to amplify weak signals before they travel down the coaxial line. This reduces transmission line losses and enhances overall receive sensitivity.

4. Signal processing, noise reduction, and filtering may be improved using Digital Signal Processing (DSP) methods and Software-Defined Radios (SDRs). DSP enables real-time enhanced noise filtering and signal augmentation.

5. Use adaptive polarization systems to automatically modify the polarization of broadcast and received signals to maximize communication depending on real-time circumstances.

6. Maintain perfect frequency control and timing accuracy. EME communication requires frequency stability, and maintaining synchronization may improve signal coherence.

7. Use sophisticated tracking technologies that alter the antenna's location in real-time to maintain it pointing toward the moon. This adjusts for the rotation of the Earth and the movement of the moon, guaranteeing a steady and dependable communication connection.

8. Recognize and correct for the path loss caused by moonbounce. To counteract signal attenuation during the reflection phase, use greater broadcast power, bigger antennas, and better modulation methods.

9. Be mindful of moon noise, which is thermal noise caused by the moon. To reduce the influence of moon noise on weak signals, consider utilizing lower noise temperature receiving equipment and customizing your setup.

10. Coordinate EME schedules with other operators to increase the likelihood of effective communication. The coordination of equipment, frequencies, and timing increases the likelihood of a successful moonbounce.

11. Consider geography, nearby noise sources, and other issues while locating your station. A well-chosen QRA locator may considerably increase the likelihood of successful EME transmission.

Frequently Asked Questions

1. How do you set up for EME contacts?
2. What are the challenges and solutions in transverter use?
3. How do you enhance band coverage with transverters?
4. How do you build and modify microwave equipment?

CHAPTER THIRTEEN
HAM RADIO CONTESTING STRATEGIES

Overview

Chapter thirteen brings us to discussing Ham Radio contesting strategies such as advanced contesting techniques in Ham Radio and so much more.

Advanced Contesting Techniques

Multi-Operator Contesting Stations

Contesting is a popular part of amateur radio in which the goal is to make as many connections as possible in a limited amount of time, generally under strict rules and regulations. Multi-operator stations improve the competing experience by enabling a group of operators to work together to maximize their performance.

Here are some important characteristics of multi-operator contesting stations:

Team Structure

- **Operator tasks:** Different operators may take on specialized tasks, such as operating stations (making continuous connections) or searching and pouncing (looking for new stations to contact).
- **Logger**: A logger is someone who maintains track of contacts, exchanges, and other relevant information. This job is critical for eliminating repeated interactions and submitting correct logs.

Station Configuration

- **Multiple Transceivers:** Most multi-operator stations feature numerous transceivers, each with its antenna. This enables simultaneous operation on many bands or modes.
- **Antenna Farm**: An array of antennas is often used to cover many bands and orientations. The selection of an antenna is crucial for improving signal strength and avoiding interference.

Coordination and communication

- **Headphones and intercoms:** Operators use headphones to focus on their jobs, while intercom systems enhance team communication.
- **Team Coordination:** During the contest, effective communication is critical for coordinating frequency adjustments, monitoring the log, and making strategic choices.

Logging Software

- **Networked Logging:** Contesting stations often utilize logging software that enables real-time cooperation and data exchange among operators. This is particularly significant in situations with several operators.

Contest Strategy

- **Band and Mode Planning:** Teams strategize by selecting when and where to operate on certain bands and modes. This entails forecasting propagation circumstances and maximizing available resources.
- **Breaks and Shifts:** Since managing operator tiredness is critical, teams plan breaks and shift changes to ensure a high level of performance throughout the competition.

Submissions and Scoring

- **QSO Points:** According to the contest regulations, stations get points for each successful contact (QSO).
- **Multipliers:** To get multiplier points, contestants must operate stations in various geographic locations or with specified qualities.
- **Log Submission:** Following the contest, the team submits their log for a score to the event organizers.

Station Maintenance

- **Maintenance:** Teams may be required to do maintenance throughout the competition to fix technical concerns and ensure that all equipment functions properly.

Contesting and Propagation Predictions

Utilizing Propagation Software for Contesting

Propagation software is a very useful tool for amateur radio (ham radio) operators, particularly when it comes to contesting. In ham radio, contesting entails making as many connections as possible in a certain time frame. Understanding and using propagation software may assist operators in increasing their odds of effective communication.

Understanding Propagation Conditions

- Propagation software offers data on the present ionosphere and other atmospheric variables. It explains how radio waves travel at a certain time and frequency.
- Discover how different propagation modes, such as ground wave, sky wave, and sporadic-E, affect signal propagation.

Choosing the Best Frequencies

- Propagation software may offer suitable communication frequencies depending on current circumstances. This is especially important during competitions, as effective frequency selection may boost your chances of creating successful interactions.
- In the program, look for frequency forecast charts or real-time propagation statistics to select bands with good circumstances.

Monitoring Band Openings

- Keep an eye on the program for real-time band opening updates. Changes in ionospheric circumstances may cause some bands to open up, allowing for greater communication over longer distances.
- Use the program to estimate when and where vacancies may appear, providing you with a competitive edge.

Antenna Orientation and Beam Direction

- Propagation software often contains tools that assist you in determining the best direction for your antenna. Changing the beam direction of your antenna depending on propagation circumstances may greatly enhance signal strength and dependability.

- Search for tools that offer information on beam headings for certain areas or regions.

Using Propagation Beacons

- Some propagation software contains data on propagation beacons, which emit signals on certain frequencies. Monitoring these beacons may provide you with real-time information on propagation circumstances.
- Assess the quality of the propagation path using the beacon information and change your operating frequency and mode appropriately.

Logging and QSO Administration

- Some complex propagation software may work in conjunction with logging software. This integration may help to speed up the process of tracking contacts during a contest.
- Check that the propagation program is compatible with your logging software, or use the propagation software's functionality for basic contact recording.

Predicting and Adapting to Changing Conditions

Understanding and coping with many aspects that might impact radio communication is required when predicting and adjusting to changing situations in ham radio.

Here are some important considerations:

Propagation Conditions

- **Prediction Tools**: Predict ionospheric conditions and radio propagation using online tools and software. Forecasts based on solar and geomagnetic activity can be provided by tools such as VOACAP (Voice of America Coverage Analysis Program).
- **Real-time Monitoring**: Use tools like WSPR (Weak Signal Propagation Reporter) and reverse beacon networks to monitor real-time propagation conditions. These services might assist you in observing the propagation of your signals.

Solar Activity

- **Solar Flux and Sunspots**: Solar activity has a significant impact on ionospheric conditions. Solar flux and sunspot counts should be monitored as indications of ionospheric conditions. Higher solar activity promotes greater HF propagation in general.

Frequency Planning

- **Band Selection:** Different HF bands may be better appropriate for communication depending on the time of day and current ionospheric conditions. Maintain a flexible approach to band selection in response to changing situations.
- **Antenna Tuning**: Adapt to changing impedance circumstances with antenna tuners and adjust your antenna system for current frequency and propagation characteristics.

Weather Conditions

- **Local Weather**: Weather may affect VHF and UHF propagation. Signal transmission can be hampered by rain, humidity, and temperature inversions. Keep up to date on local weather conditions.
- **Emergency Preparedness**: Be ready for inclement weather by having backup power, waterproofing equipment, and emergency communication procedures in place.

Geothermal Activity

- **Monitoring the K-Index:** Keep a watch on the geomagnetic K-index. Higher K-index values imply higher geomagnetic activity, which may affect HF radio transmission. K-index data is provided in real-time by space weather forecast centers.

Frequently Asked Questions

1. How do you predict and adapt to changing conditions?
2. How do you use propagation software for contesting in Ham Radio?
3. What are the different Ham Radio contesting strategies?

CHAPTER FOURTEEN
HOMEBREWING AND DIY PROJECTS

Overview

In this chapter, you will learn about homebrewing and DIY projects in Ham Radio including building high-gain yagi antennas.

Advanced Antenna Constructions

Building a Quad Antenna

Without question, the greatest DX antenna available is the cubical quad antenna. It has a better gain than any other antenna of comparable size and is also relatively inexpensive to build. At heights less than one wavelength, one of the most significant benefits of the quad is that it has a smaller angle of radiation than a yagi. This makes it an excellent antenna for working skip since more power is emitted at low angles, even if they cannot be put as high as desired.

One of the reasons for the quad's efficacy is that it not only has a higher forward gain but also a bigger physical capture area than a yagi with the same boom length. Each element has a full wavelength in size, as opposed to a yagi's half-wavelength elements, making the quad a significantly more effective antenna for both TX and RX. As with any antenna, real performance is only obtained when all components are resonant at their proper frequencies and the feed point is properly matched. The figure below depicts the fundamental structural design and measurements of a typical four-element cubical quad.

These dimensions must be followed as precisely as possible. The Driven element is tuned to resonance in the center of the band for which it was created. Once the element has been trimmed to the proper size, the feedline may be joined and carefully adjusted using a dip meter or an antenna analyzer.

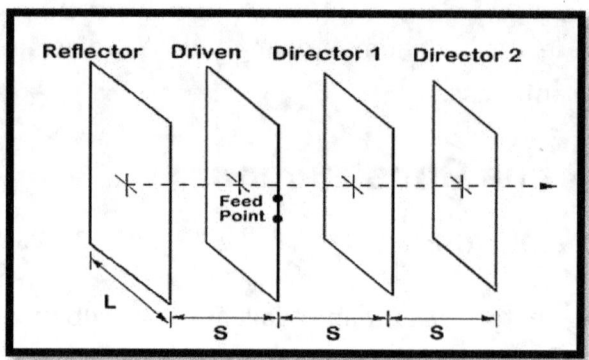

Quad Dimensions

The following are the formulas for computing quad-element lengths (in centimeters):

- 78.5/Frequency (Mhz) = Reflector Element (L).
- 74.5/Frequency (Mhz) Driven Element (L)
- Elements of Direction (L) = 70.8/Frequency (Mhz)

If you are without a dip meter, once all the components are in place, the size of the driven element may be modified to get the lowest SWR. The reflector element is set to resonance at a 5% lower frequency than the driven element, while the directors are both tuned to resonance at a 5% higher frequency than the driven element. Again, a dip meter comes in helpful to ensure that all of the components are tuned to their respective frequencies. If the resonant or tuned frequency of any of the parasitic or "non-driven" components approaches that of the driven element, the SWR increases dramatically and the gain decreases significantly. This is why it is vital to precisely alter the length of each element to get the best gain and lowest SWR. To achieve a low SWR throughout the whole band on bands with a high bandwidth (such as 10 meters), the parasitic components may be adjusted up to 7% away from the driving element. This will function OK, however, the maximum forward gain will be somewhat reduced to get the broader bandwidth.

The quad calculator is an extremely handy tool that not only calculates element size but also determines the optimal element spacing for optimum gain and provides spreader arm lengths. Do not forget that this calculator is meant to provide maximum gain and does not account for feed point impedance. To feed directly using 50 Ohm coax and a 1:1 balun, the distance between components should be maintained at roughly 0.12 wavelengths. If you're using an impedance-matching device, these dimensions are appropriate and will give you the most gain.

Quad Antenna Feed Systems

The matching system is one of the most important features of the quad design after all of the pieces have been tweaked and sorted. The matching system's goal is to create a near-impedance match between the feedline and the driven element feed point. The feed-point impedance changes with the distance between the driving and parasitic components. At 0.15 wavelength spacing or less, the impedance is extremely similar to 50 Ohms and may therefore be supplied directly with a 50 Ohm cable. The effects on the feed-point impedance reduce as the element spacing grows, and the impedance climbs closer to what it would be as a single element at roughly 120-140 ohms. A quarter wave 70-ohm matching stub is often used at this larger spacing. The stub is about 0.66 of the physical length and is an electrical quarter wavelength rather than a physical wavelength.

At the larger element spacing, this matching stub offers a tight match to the impedance of the driven element. In addition to matching the impedance, we must ensure that no current flows through the coax's outer shield. This is often a problem when connecting an unbalanced coaxial cable to a balanced antenna. The solution to the RF flowing on the outside shield of the coax is to utilize some form of balun. A balun is a device that converts unbalanced line currents to balanced line currents, resulting in no or very little RF on the coax shield. The simplest basic balun is made by looping three twists of coax through an iron ferrite core one electrical quarter wavelength from the feed point. We already have the maximum impedance point at an electrical distance of one-quarter wavelength from the feed point, so by adding a balun here, we produce even greater impedance, thereby stopping any RF on the braid from traveling down the coaxial.

Covering the cable with a metallic sleeve one electrical quarter long is an even simpler kind of balun that also provides very high impedance on the outside braid. This sleeve is connected to the coax's outer braid at one electrical quarter wavelength from the feed point. This is a simple solution that prevents stray RF from traveling on the coax shield. Pay great attention to the matching system and balun, since they will ultimately decide the antenna's overall performance. Without a balun, the radiation pattern and F/B ratio are often suboptimal.

Building High-Gain Yagi Antenna in Ham Radio

Materials Required

1. **Boom**: A non-conductive element support framework. This is often made of a strong, lightweight material such as PVC pipe or fiberglass.
2. **Driven Element**: The radio feedline-connected active element.
3. **Reflector**: The element that is located behind the driving element and reflects radio waves forward.
4. **Directors**: Elements placed in front of the driving element to assist in concentrating the radio waves.

Tools Required

1. **Measuring Tape**: For precise element length measurements.
2. **Wire**: For the elements, use copper or aluminum wire.
3. **Coaxial Cable**: To connect the driven element to the radio, use a coaxial cable.
4. **Soldering Iron**: For connecting components and coaxial cable, use a soldering iron.
5. **Drill**: To make holes for mounting elements.
6. **Connector**: This is used to connect the coaxial wire to the radio.

Procedure

1. Determine the frequency for your Yagi antenna. The size of the components is determined by the frequency at which you want to operate. Calculate the lengths of the driving element, reflector, and directors using an online Yagi antenna calculator. These calculators will also include parameter information such as element spacing and other factors.
2. Cut the driving element, reflector, and directors from the wire using the calculator measurements. Also, cut the parts precisely to the predicted lengths.
3. Connect the coaxial wire to the driving element. Solder the connection and thoroughly insulate it. Place the reflector behind the driving element and the

directors ahead of it. Maintain correct spacing according to the calculator's recommendations. Use insulators or clamps to secure the elements of the boom.

4. For best performance, the Yagi may need to be calibrated. You may do this by altering the lengths of the parts slightly. To discover the optimal fit, use an antenna analyzer or an SWR meter attached to your radio.

5. To enhance its efficacy, place the Yagi antenna in a position with a clear line of sight. Consider using a mast or pole to elevate yourself.

6. Use your radio equipment to test the Yagi antenna. Check the SWR and make any necessary modifications.

Transmitter and Receiver Design

Understanding RF Circuitry for Hams

Radiofrequency (RF) circuits are electrical devices that USE RF signals to communicate between devices. An RF circuit's principal duties are to receive signals from an antenna and to broadcast them to another device.

What exactly is an RF Circuit?

An RF circuit (radio frequency circuit) is an electrical device that employs radio frequency signals to transport data between devices. An RF circuit's principal duties are to receive signals from an antenna and to broadcast them to another device. Wireless communication equipment often uses RF circuits. Cell phones and Wi-Fi routers, for example, USE RF circuits to relay messages between devices. RF circuits are often designed to operate at very high frequencies. This implies that their signals move at high speeds (often between 300 MHz and 30 GHz). This is why, when you hold a gadget close to an RF circuit, you can only hear static.

What Is the Function of an RF Circuit?

The receiving channel and the sending channel comprise the RF chip architecture. The transmitter is the circuit component that generates the electromagnetic wave. The receiver is the component of the circuit that converts the wave into an electrical signal. The antenna receives a signal, which is then passed via the transmitter to be converted into an electromagnetic wave. Finally, the signal passes through the receiver and is converted back to an electrical signal. The receiver is often linked to another device (such as a speaker) so that the information may be heard or displayed. The receiving circuit consists of an antenna, an antenna switch, a filter, a low-noise amplifier, a receiver demodulator, and other components. The receiver first converts the electromagnetic wave into an electrical current. The electrical current then passes through the filter and amplifier to produce a feedback signal. The feedback signal is then transferred to a demodulator, which converts it into baseband information such as RXI-P, RXI-N, RXQ-P, and RXQ-N before being delivered to the logic audio circuit for further processing.

Transmitting Circuit

An RF transmitter's principal job is to convert your data into an electromagnetic wave that can be sent to another device via an antenna. Electricity is used by an RF transmitter to generate electromagnetic waves.

RF Transmitter Structure

The voltage-controlled oscillator (VCO), amplifier, phase detector, and modulator are the basic components of an RF transmitter. The oscillator generates alternating current at a fixed frequency determined by the number of oscillations per second. This is referred to as the carrier frequency. The signal is amplified by the amplifier before it is delivered via the antenna. The modulator is used to control the amplitude, frequency, and phase of the carrier wave, allowing data to be transmitted.

Working Methods

The radio frequency (RF) transmitter generates an electromagnetic signal in the electromagnetic spectrum when coupled to an antenna. It operates by creating a carrier wave at the desired frequency and modulating it with an input signal such as speech or music. The carrier wave is then amplified and broadcast via the antenna. The carrier wave is broadcast by the antenna, and the signal is picked up by the receiver. The signal is then demodulated and the data is extracted by the receiver. This is referred to as modulation and demodulation.

RF Circuit and their Importance

Any electrical circuit designed to operate at radio frequencies is referred to as an RF circuit. Many consumer and industrial products, such as remote controls, use RF circuits. Wireless networks, such as WiFi networks, also make use of RF circuits. A circuit diagram or a schematic diagram is a diagram of an RF circuit. An RF circuit diagram depicts the electrical components and their connections in the circuit. Engineers must first understand the characteristics of radio frequencies, such as impedance and radio wave propagation, before designing an RF circuit. Engineers must also be familiar with the design tools used to create RF circuits, such as circuit analysis and circuit simulator software. Because RF circuits are designed to operate at high frequencies, they must be adequately shielded to avoid interference with nearby circuits. To avoid interference with other electrical systems in buildings, RF circuits must be properly grounded.

RF Circuits' Common Applications

Radio frequency (RF) circuits are commonly found in wireless communication devices such as radios and cell phones. Circuits are typically used to transmit data over short distances between devices. A WiFi router, for example, may use an RF circuit to send and receive data from nearby computers. By bouncing off a satellite, RF circuits can also be used in long-distance communication. For example, while in space, ham radio operators use RF circuits to send and receive data. These are the most common RF circuit-based systems or devices.

Wireless communication networks

- Radio transmitters and receivers
- Satellite communication
- Broadcast television systems
- Radios and cell phones

Automotive applications

- GPS and navigation systems
- Automotive collision avoidance systems
- Navigation systems

Radar systems

- Distance monitoring
- Antenna systems
- Remote sensing systems

Smart home systems

- Electricity meters
- Household gas meters
- Surveillance cameras

Building Transmitters and Receivers from Scratch

The steps:

1. Become acquainted with electronic components, circuits, and soldering techniques. Know about resistor codes, capacitor values, and other fundamental electronic principles.
2. Learn about radio frequency (RF) theory, which includes topics such as frequency, wavelength, modulation, and antenna theory. The ARRL Handbook and other ham radio literature can be beneficial.
3. Select the frequency and band on which you want your transmitter to operate. Make sure you follow the rules that govern amateur radio operations in your area.
4. Choose a modulation strategy for your transmitter. Amplitude modulation (AM), frequency modulation (FM), and single sideband (SSB) are often used in amateur radio.
5. Create the transmitter circuit depending on the frequency, band, and modulation scheme you've selected. This might include creating oscillators, amplifiers, and modulation circuits. If available, use simulation tools.
6. Put the circuit together on a breadboard or bespoke PCB. To guarantee adequate operation, test each piece of the circuit. Use oscilloscopes and frequency counters for testing.
7. If your transmitter demands additional power, a power amplifier stage may be required. Consider impedance matching and heat dissipation.
8. Include safety precautions such as fuses and adequate grounding to safeguard yourself and your equipment.

Building a Receiver

Here are the steps:

1. Check several receiver designs, including superheterodyne, direct conversion, and regenerative. Understand the benefits and drawbacks of each.
2. Select the frequency and band for your receiver in the same way you did for the transmitter. Ensure that all requirements are followed.
3. Create the receiver circuit, which will include the front end, mixer, local oscillator, intermediate frequency (IF), and audio stages. Consider applying selectivity filters.
4. Assemble the receiver circuit and test each step to confirm that it works properly. Make use of an antenna and tuning circuitry.
5. Consider adding AGC to automatically change the receiver's gain to maintain a steady output level.
6. Add an audio amplifier to power a speaker or headphones. Check that the audio is clean and free of distortion.
7. Ensure you add safety precautions in the reception circuit, just as you did in the transmitter.

Frequently Asked Questions

1. How do you build a Quad antenna?
2. How do you build transmitters and receivers from scratch?
3. How do you build high-gain Yagi antenna in Ham Radio?
4. How do you understand RF circuitry for Hams?

CHAPTER FIFTEEN
RADIO FREQUENCY INTERFERENCE (RFI) MITIGATION

Overview

We have come to the last chapter. Here, we will talk about radio frequency interference mitigation, the different techniques for RFI detection, and others.

Identifying and Location RFI Sources

Categories of Interference

In general, interference can be broken down into two categories: narrow-band interference and broadband interference.

- **Narrow Band**: This would include modulated continuous wave (CW) transmissions as well as continuous wave (CW) communications. Among the many examples that may be given are clock harmonics from digital devices, co-channel broadcasts, adjacent-channel transmissions, intermodulation products, and so on. On a spectrum analyzer, this would seem to be thin vertical lines or somewhat broader modulated vertical bands associated with certain frequencies. Both would be associated with the same frequencies.

- **Broadband**; this would largely comprise switch-mode power supply harmonics, arcing in overhead power lines (power line noise), wireless digitally modulated systems (like Wi-Fi or Bluetooth), or digital television. Broadband would also include power line noise. This would seem to be a wide range of signals or a rise in the noise floor when seen through the lens of a spectrum analyzer. Among the most typical causes, power line noise and switch-mode power supply are often encountered.

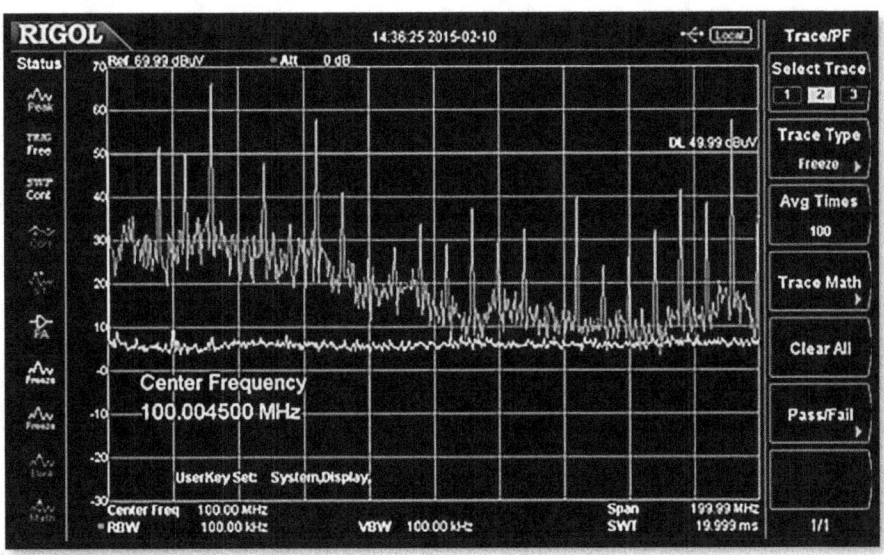

Types of Interference

In the following paragraphs, we will discuss some of the most typical forms of interference.

- **Co-Channel Interference:** The phenomenon known as co-channel interference occurs when many transmitters (or digital harmonics) use the same receive channel or fall into the same received channel.
- **Adjacent-Channel Interference**: A transmitter that operates on a neighboring frequency and whose energy spills over into the adjacent channel is referred to as adjacent-channel interference.
- **Intermodulation-Based Interference**: The phenomenon known as intermodulation-based interference takes place when the energy from two or more transmitters combines to generate spurious frequencies that are sent to the channel that wants to receive them. Third-order mixing products are the most prevalent, and in most cases, this activity is caused by transmitters that are located nearby. The presence of a powerful signal in an area that is used for FM broadcasting is an example of the possibility of intermodulation.
- **Fundamental Receiver Overload:** The fundamental receiver overload is often brought on by a powerful transmitter that is located close to the receiver. This overloading of the receiver front-end or other circuitry may result in interference

or even suppression of the normal signal that is received. Paging transmitters that use the VHF frequency range may often cause interference with receivers.

- **Power Line Noise (PLN):** Power Line Noise (PLN) is a broadband interference issue that is rather prevalent and is often generated by arcing on electric power lines and related utility devices. There is a loud, scratchy buzzing sound that may be heard in an AM receiver. The interference can spread into the high-frequency spectrum, depending on the closeness to the source, and it may begin at extremely low frequencies that are below the AM broadcast range. It can extend up into the UHF spectrum if it is found near enough to the source.
- **Switch-Mode Power Supplies**: Switch-mode power supplies are quite widespread and are used for a wide range of consumer or business items. They are also a typical cause of interference for broadband networks. Another significant source of interference is lighting equipment, such as the more recent LED-based lights or the commercial agricultural "grow" lights from the agricultural sector.

The following are some examples of transmitters that are regularly responsible for radio frequency interference (RFI):

- **Two-way or Land Mobile Radio:** Strong FM frequencies that interfere with two-way or land mobile radio transmissions might cause a phenomenon known as "**capture effect**," which is the overriding of the signal that is wanted to be received.
- **Paging Transmitters**: Paging transmitters are often exceptionally strong FM or digitally modulated broadcasts that have the potential to overwhelm receivers. There is a possibility that digital paging could interfere with a broad variety of receive frequencies and will have a highly scratchy sound, similar to that of a power saw or buzzing. One fortunate development is that the majority of the VHF paging transmitters has relocated to the 929/931 MHz frequency pairs, which means that this problem is no longer an issue.
- **Broadcast Transmitters:** Interference from broadcast transmitters will have modulation characteristics that are comparable to those of their broadcasts, whether they are AM, FM, video carriers, or digital signals.

Cable Television: Generally speaking, signal leakage from cable television systems will take place on the channels that are assigned to them according to their defined schedules.

Many of these channels are redundant with other over-the-air radio communications channels that are already in existence. Interference may be comparable to wideband noise if the signal that is leaking is a digital channel. This is because a digital cable channel is nearly 6 MHz wide.

Wireless Network Interference: Interference to wireless networks (such as Bluetooth, Wi-Fi, and others) is becoming more widespread. Due to the expansion of mobile, home (Internet of Things), and medical devices that include Wi-Fi and other wireless modes, this problem is anticipated to become much more severe.

Locating RFI

Simple Direction Finding (DFING)

DF Techniques: Techniques for DFing comprises two basic approaches to the process. (1) The "**Pan 'N Scan**" technique, in which you "pan" a directional antenna and "scan" for the signal that is interfering with your signal, marking the direction on a map while making a note of any areas where the lines overlap. Secondly, "Hot and Cold," which involves the use of an omnidirectional antenna while simultaneously monitoring the signal intensity. When using this technique, the rule of thumb is that for every six-decibel shift, you have either doubled or halved the distance to the source of interference for the signal. For instance, if the signal intensity was -30 dBm at a distance of one mile from the source, then the spectrum analyzer should read around -24 dBm while going to a distance of forty-five meters from the source.

DF Systems: Radio direction-finding (RDFing) equipment comes in the form of DF Systems, which can be placed inside a vehicle or utilized as portable equipment. Many automatic Doppler direction-finding technologies are available for use in vehicle conditions.

Step Attenuator: During the process of DFing, you will also discover that a step attenuator is quite useful. As you go closer to the interference source, this gives you the ability to exercise control over the signal strength indicator (especially the receiver overload). Those versions that are considered to be the best have a range of at least 80 decibels and come in stages of 10 decibels. You may get step attenuators by purchasing them from electronic wholesalers like DigiKey and other such companies. This would include commercial suppliers such as Narda Microwave, Fairview Microwave, Arrow, and other such companies.

Locating Power Line Interference

- Low-Frequency Interference: The interference path may include radiation that is caused by conducted emissions along power lines when it comes to low-frequency interference, and this is especially true for power line noise (PLN). Therefore, while you are employing the "**Hot and Cold**" approach, you will need to be aware that the radiated noise will typically follow the path of the power lines, reaching its highest point along the path and then falling to its lowest point along the path. In most cases, the greatest peak is indicative of the true source of the noise. There may be several sources of noise, some of which may be located at a considerable distance away.

- **Antennas**: The built-in "**loopstick**" antenna on an AM broadcast band radio or the telescoping antenna on a shortwave radio may be effective for listening to power line noise. Both of these antennas are suitable for listening to power line noise. However, you will want to use higher frequencies to trace down power line noise to the source pole, and in most cases, you will also want to use them to DF other sources that are interfering. This form of broadband radio frequency interference (RFI) may be effectively received by a simple directional Yagi antenna, such as the Arrow II 146-4BP with three-piece boom (www.arrowantennas.com), which can be erected in a short amount of time and connected to a short length of pipe.

- **Use of VHF Receivers**: When it comes to DFing, you should normally try to make use of VHF or higher frequencies whenever it is available. The shorter wavelengths not only increase the accuracy with which the source can be located but also make it more feasible to use smaller portable antennas.

- **Signature Analyzers:** Signature analyzers are time-domain interference-locating devices that generate a unique "**signature**" of an interference signal. These analyzers are useful for identifying interference in the time domain. Instruments that are manufactured by Radar Engineers would fall under this category. When it comes to locating power line noise and consumer products that generate repeated noise bursts with a defined frequency, they are the most effective options.

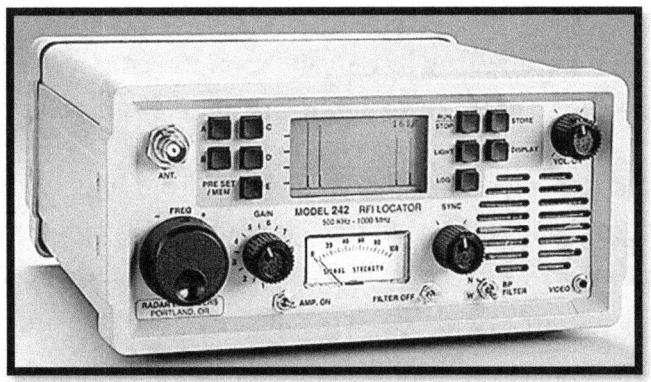

Locating narrow-band interference

The spectrum analyzer is the instrument that is suggested for the majority of narrow-band interference sources, such as co-channel, neighboring channel, and intermodulation interference. This is because the spectrum analyzer enables you to concentrate on certain frequency channels or bands while simultaneously providing a comprehensive and comprehensive view of the situation. After the signal that is causing interference has been discovered, the analyzer may be used to perform DF on the signal.

Using Spectrum Analyzers

Frequency versus amplitude of radio frequency (RF) transmissions is shown by spectrum analyzers. They can be useful in distinguishing the kind of interference and the frequencies of the signals that are interfering, particularly in the case of narrowband interference. Real-time analyzers and swept-tuned analyzers are the two categories of analyzers. Swept-tuned analyzers can show a chosen bandwidth from start to stop frequencies. They are based on a superheterodyne concept and use a tunable local oscillator so that they may display the required bandwidth. Due to the extended sweep time, they are effective in showing signals that are continuous or almost constant; nevertheless, they struggle to capture signals that are transient and intermittent. To do an analysis of the spectrum that has been acquired, a real-time analyzer takes a sample of the spectrum and applies digital signal processing methods to it. When it comes to recognizing and finding signals that may not even show up on sweep analyzers, they are useful since they can record transient intermittent signals and are used for this purpose. The majority of real-time bandwidths are restricted to speeds of no more than 27 to 500 megahertz. Both the Signal Hound BB60C and the Tektronix RSA306 are real-time spectrum analyzers that are powered by USB and communicate with a personal computer for control and display. Both of these analyzers are quite affordable.

Considering that spectrum analyzers have an untuned front end, it is essential to bear in mind that they are especially vulnerable to high-powered neighboring transmitters that are off-frequency from where you may be searching. This is a crucial factor to keep in mind while using spectrum analyzers. This can result in the production of internal intermodulation products, also known as spurious responses, or erroneous amplitude measurements, which may be very deceptive. When employing spectrum analyzers in an environment that is "RF rich," it is essential to make use of bandpass filters or tuned cavities (duplexers, for example) at the frequency that is of interest. In addition, spectrum analyzers help determine the characteristics of commercial broadcast, wireless and land mobile communications systems. Real-time analyzers are the most effective tools for dealing with wireless or intermittent interference. When the analyzer is being used for tracking PLN, it is recommended that it be placed in the "**zero-span**" mode so that the amplitude fluctuation may be seen. In addition, putting the analyzer in the "Line Sync" mode can be of use.

Commercial Interference Hunting Systems

Several different manufacturers deal with interference-hunting or direction-finding systems. As was mentioned earlier, a real-time spectrum analyzer is the most effective instrument for capturing brief, intermittent signals, some of which can be as short as a few microseconds. This is especially true for digitally modulated signals or intermittent interference, which is especially prevalent in commercial communications installations. For instance, the Aaronia Spectran V5 series may be considered an example. either the Narda IDA2 or the Tektronix RSA-series.

- **Aaronia**: The Aaronia is not only the most portable and lightweight device for DFing, but it also seems to be the largest and heaviest of all the systems. Spectran V5 Handheld is the company's real-time analyzer that is the smallest in size. There is no mapping capability available on this model; however, the bigger Spectran V5 XFR PRO is a ruggedized laptop that is capable of using open-source maps and includes capabilities that allow for triangulation. Additionally, Aaronia offers a selection of reasonably priced directional antennas, and some versions may provide the option to add a combined GPS and compass.

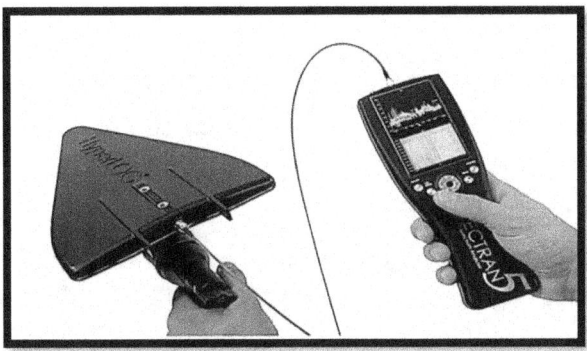

The fact that Aaronia has created a drone detection system that is made of a 3D tracking antenna, the model IsoLOG 3D, with choices ranging from 9 kHz to 40 GHz in 360 degrees is another thing that sets them apart from their competitors. This corresponds to their Spectran Command Center, which has three LCD panels in several displays. Take a look at the references for further details on that system.

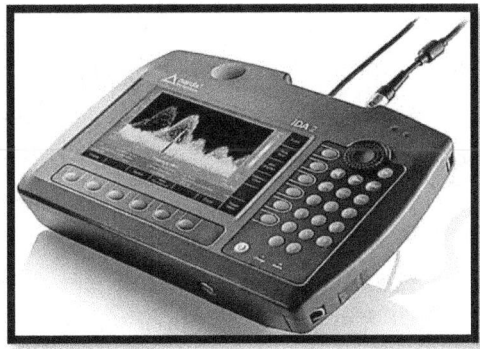

RFI Elimination Techniques

Interference with radio frequencies, often known as RFI, is a problem that frequently arises in the field of ham radio operations. This interference can be generated by a wide variety of sources, including electrical equipment, power lines, and other radio transmitters.

On the subject of ham radio sets, the following are some ways that might assist in reducing or limiting RFI:

Ferrite Chokes and Beads

- **Placement**: Attach ferrite chokes or beads to the feedline, power cords, and any other cables that are connected to your radio equipment. This helps to minimize common-mode currents, which may contribute to radio frequency interference (RFI).
- The number of turns is the number of times that the cable is wound through the ferrite. This will boost the efficacy of the cable.

Baluns

- In the case of feedlines and antennas, the use of baluns, which are balanced to unbalanced transformers, might be of assistance in addressing common-mode currents. Be careful to choose baluns of a good grade that are suitable for the frequency range you are working with.

Proper Grounding

- You must make certain that your station is equipped with a reliable grounding system. Grounding in the correct manner may assist in the dissipation of unwanted radio frequency energy and lessen the possibility of interference.

Isolation

- Maintaining a physical separation between your radio equipment, power sources, and other electrical devices is an important part of the separate systems strategy. Because of this, the likelihood of interference is reduced.

Filters

- It is recommended that you install bandpass filters on the feedline of your antenna to lessen the influence of signals that are outside of the band. If you are encountering interference from adjacent transmitters operating on various frequencies, this is a really helpful feature.

Shielded Cables

- **High-Quality Cables:** Use shielded cables of the highest possible quality for your connections. Since this is the case, RF from the outside can be prevented from entering your system.

Antenna Tuner

- **Use of an Antenna Tuner**: A decent antenna tuner may assist in matching your antenna system to your transceiver, hence lowering the likelihood of reflections and reducing the amount of radio frequency interference (RFI).

Power Line Filters

- Use power line filters on the AC power lines that are linked to your equipment. This is the first step in installing filters. This has the potential to lessen the amount of interference.

RFI Sniffer

- Use radio frequency interference (RFI) sniffer or a directional antenna to find the source of interference before proceeding with the next step. As soon as the issue has been detected, you will be able to take precise actions to address it.

Frequently Asked Questions

1. How do you locate RFI in Ham Radio?
2. What are the different RFI elimination techniques?
3. How do you identify RFI sources?

CONCLUSION

Ham Radio is a popular and common hobby and service that uses specified radio frequencies for the purpose of non-commercial message exchange, wireless experimentation, self-training, and emergency communications. Amateur radio is the only service that is controlled by international convention, and that is amateur radio. Over the course of its history, ham radio has often proved its capacity for resilience and flexibility. The community of amateur radio operators throughout the world has continued to expand and develop, welcoming new technology while maintaining the fundamental principles of communication, experimentation, and public service. Hams have probably investigated new technologies like software-defined radios, digital modes, and satellite communications to improve their capabilities as technology has advanced.

It is probable that the devotion of the amateur radio community to public service and emergency communication is still strong, as operators continue to provide their expertise and equipment during times of crisis. Additionally, it is probable that the feeling of camaraderie and the sharing of information that exists within the ham radio community will continue to exist, which will help to cultivate an atmosphere that encourages collaboration among enthusiasts of all ages. If you want to keep up to date on the most recent advancements and trends in the field of amateur radio, it is very necessary to maintain connections with important ham radio periodicals, online forums, and community events. You can broadcast radio communications on a variety of frequency bands that have been designated exclusively for radio amateurs if you are certified as a radio amateur.

INDEX

F

G

H

I

O

P

Q

R

U

V

W

X

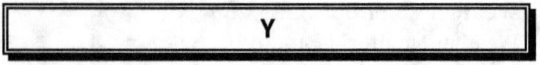

Y

Yaesu FT-60R, 56

www.ingramcontent.com/pod-product-compliance
Lightning Source LLC
Chambersburg PA
CBHW082208290526
45794CB00009B/3472